User Guide to the Bibliography and Index of Geology

Edited by John Mulvihill

 The American Geological Institute

USER GUIDE TO THE BIBLIOGRAPHY AND INDEX OF GEOLOGY

Contents

There is a growing awareness of the vital importance of information about the Earth. Energy sources, strategic minerals, earthquake prediction, volcanic eruptions, nuclear reactor sites, water pollution, landslides, erosion . . . . These problems present geoscientists, students, and teachers with unanswered questions.

Thousands of technical papers reporting on these problems are published every year. They are in many languages, from many countries. Organizing this information so that it can be made available to all requires an information system. To meet that need, the American Geological Institute provides the Bibliography and Index of Geology. This manual is a guide to the Bibliography from 1969 to the present. If you need more specific help, ask your librarian. Or, in the U.S. (outside of Virginia), you can call us toll-free on 800/336-4764, or write to the GeoRef Information System, American Geological Institute, 5205 Leesburg Pike, Falls Church, Va., 22041.

1. SCANNING AND SEARCHING

The Earth Science Literature

The earth-science literature is an ongoing, evergrowing record of research. Journals containing reports of current research provide the basic records. In addition, many other "primary" sources of information are important. These include materials presented at scientific meetings, government reports, books, maps, and monographs.

The volume of material has grown until finding information has become a major research problem. The American Geological Institute, working from the primary documents, classifies and indexes the information in such a way that it can be found—now and in the future. To do this, each year AGI processes over 50,000 items in 44 languages, from countries all over the world. The Bibliography and Index of Geology covers the Earth.

Continuing a Bibliographic Tradition

In 1969 the Geological Society of America published the first volume entitled Bibliography and Index of Geology. However, its volume number was 33 because this Bibliography was a continuation of the Geological Society of America's Bibliography and Index of Geology Exclusive of North America, published 1934–1968, volumes 1–32.

Coverage in the 1969 Bibliography was worldwide, as distinct from the volumes which preceded it in the GSA series. GSA and the U.S. Geological Survey had agreed that the GSA Bibliography would become worldwide and that it would replace and continue the USGS Bibliography of North American Geology, which had been published since 1923. Accordingly, the Bibliography of North American Geology ceased in 1970.

The American Geological Institute became involved with the Bibliography in 1967 when AGI began GeoRef, a computerized data base, and began to produce the Bibliography from GeoRef for GSA. As of this time the Bibliography began to be photocomposed. In 1979, AGI assumed full production and publication of the Bibliography.

The GeoRef Information System staff, with computer assistance, strives to maintain the standards of excellence that have made these geoscience bibliographies outstanding. The indexers are full time, have degrees in geology and know foreign languages. The specifications of the UNISIST Reference Manual for Machine-Readable Bibliographic Data are adhered to. Data entry and correcting are online and include immediate computer checks of the presence of data elements and, where feasible, the accuracy of the data; e.g., of index terms and serial titles. The data are also copy-edited by people for accuracy and then final-edited for completeness, especially of the indexing, by the Chief Editor. Then the data are checked again when assembled in a proof of each issue of the Bibliography.

In 1982, AGI began to receive references to current publications, in machine-readable form, from Bureau de Recherches Géologiques et Minières and Centre National de la Recherche Scientifique in France. These references are produced in several European countries and include items published in Europe and other designated areas. These references are checked on receipt, index sets are added to them, and they are incorporated into the Bibliography. In return for these references, we are sending references to the French for items published in 1981 and later, in North America and other areas. Elaborate precautions are being taken to insure the uniformity and quality of the exchanged references. This exchange should improve coverage of European geology in the bibliography and increase our efficiency in its production.

Applying Information to Research

Earth scientists need information organized to serve their need for scanning current publications and the searching among large amounts of material written in the past. You can do both with the Bibliography: the monthly issues give you a timely compilation of current literature, and the annual cumulative volumes are your tools for an in-depth literature search.

SCANNING THE MONTHLY ISSUES

Whether you want an overview of research going on in the earth sciences, or want to keep up with current developments in your own specialty, the monthly issues fill the need. Scanning the Bibliography helps you:

1. Find a *group* of papers about a general subject or a specific topic.
2. Determine which of that group contains the "best" (most useful) information.

Each issue contains 4 sections: Serials, Fields of Interest, Subject Index, and Author Index. Knowing the purpose and organization of each section will help you use the Bibliography.

Serials

Approximately 70% of the citations in the Bibliography are to serials. If you are seeking a particular issue of a journal, go to the Serials section in the front of each issue. There you will find a complete listing of all periodicals cited in that issue, including the particular journal issues covered. You will find, also, that the Serials list provides more detailed information on a periodical than is given in the bibliographic citations.

We are in the process of changing the format of entries from abbreviated titles to standardized Key Titles. As a result, you will find examples of both in the list.

Old Format:

Abbreviated serial title/Serial title/ISSN/CODEN/Volume/Issue/Publication date

J. Min. Metall. Inst. Jap. Journal of the Mining and Metallurgical Institute of Japan. Tokyo. ISSN: 03694194. Vol. 97, No. 1115, 1981.

New Format:

Key Title/Publisher/ISSN/CODEN/ Volume/Issue/Publication date

American Journal of Science. New Haven, CT: Kline Geology Laboratory, Yale University, 1880–. ISSN: 00029599, CODEN: AJSCAP. Vol. 282, No. 5, May 1982.

Notes on the Serials List

Titles of foreign-language serials are cited in the original language, if that language uses the Roman alphabet. Titles in Cyrillic-alphabet languages, such as Russian, have been transliterated into Roman alphabet, but not translated. Titles in languages using neither Roman nor Cyrillic alphabet, such as Arabic or Japanese, are given in English translation.

ISSN is short for International Standard Serial Number, an identification code issued by the International Serials Data System. ISSNs usually consist of 7 numbers and a check character. The ISSN for the Annals of the Carnegie Museum, for instance, is 00974463.

CODEN is another, similar identification code for serials, usually consisting of 6 letters. The CODEN for the Annals of the Carnegie Museum is CIMUAU. CODENs have 5 letters and a check character.

Fields of Interest

The Fields of Interest section holds the bibliographic citations and fills nearly half of the book. This is the browsing section. You can scan it for an overview of what is happening in the geosciences, and also scan your special areas of interest to keep up with current publications. You will find references grouped in the following broad subject categories:

01—Mineralogy and crystallography
02—Geochemistry
03—Geochronology
04—Extraterrestrial geology
05—Petrology, igneous and metamorphic
06—Petrology, sedimentary
07—Marine geology and oceanography
08—Paleontology, general
09—Paleontology, paleobotany
10—Paleontology, invertebrate
11—Paleontology, vertebrate
12—Stratigraphy, historical geology and paleoecology
13—Areal geology, general
14—Areal geology, maps and charts
15—Miscellaneous and mathematical geology
16—Structural geology
17—Geophysics, general
18—Geophysics, solid-earth
19—Geophysics, seismology
20—Geophysics, applied
21—Hydrogeology and hydrology
22—Engineering and environmental geology
23—Surficial geology, geomorphology
24—Surficial geology, Quaternary geology
25—Surficial geology, soils
26—Economic geology, general and mining geology
27—Economic geology, metals
28—Economic geology, nonmetals
29—Economic geology, energy sources

Chapter 3 describes each of the 29 categories more fully.

Document-Type Headings

Within each category you will find references grouped alphabetically by author and numbered consecutively throughout each year under the following headings:

• *books—monographs—reports*

a. Citations for books or monographs that are not serials follow this form: citation number/author/affiliation/other authors/role/title/title in English/publisher/place of publication/pagination/illustrations/number of references/date of publication/notes

b. Citations for books or monographs that are serials follow this form: citation number/author/

affiliation/other authors/role/title/title in English/ serial title/volume/issue/pagination/illustrations/ number of references/date of publication/notes

c. Citations for reports follow the form for serial or non-serial monographs above and include a report number before the date of publication and availability information after the date.

●*meetings*

a. Citations for meeting publications that are serials follow this form: citation number/author/ affiliation/other authors/role/title/title in English/ date of meeting/place of meeting/serial title/volume/issue/pagination/illustrations/number of references/publication date/notes

b. Citations for meeting publications that are not serials follow this form: citation number/author/affiliation/other authors/role/title/title in English/date of meeting/place of meeting/publisher/ place of publication/pagination/illustrations/number of references/publication date/notes

●*dissertations and theses*

Citations follow this form: Citation number/ author/title/pagination/illustrations/number of references/date degree granted/degree-granting institution/availability/notes

●*separately published maps*

Citations follow the form for serial or non-serial monographs above and also include, in the illustration information, map type and map scale

●*individual papers and chapters*

a. Citations to papers or chapters in serials follow this form: citation number/author/affiliation/ other authors/title/serial title/volume/number/pagination/illustrations/number of references/publication date/notes.

b. Citations to papers or chapters in non-serials follow this form: citation number/author of paper of chapter/affiliation/other authors of paper of chapter/title of paper or chapter/names of monograph/author of monograph/role/publisher/ place of publication/pagination/illustrations/number of references/date/notes

For papers or chapters, the notes will include meeting information if the paper was presented at a meeting. If so, complete information on the meeting will be found in the meeting publication citation elsewhere in the Bibliography.

This subdivision by document type appears only in the monthly bibliographies. In addition to providing quick access to document types, it gives you a context for interpreting the source information found in each citation. The source information is the most complex part of a reference, because it changes with the citation's document type and bibliographic level (analytic, monographic, or collective).

Scanning for general research awareness

Your research area is seismology. Specifically you need to keep up to date on earthquakes and tectonophysics.

● Determine which of the Fields of Interest covers

these specific areas by checking the 29 subject categories. You will find earthquakes listed in categories 19 and 22, and tectonophysics in category 18.

● Scan the references under these headings in the Fields of Interest section as in the small sample given.

● Note the papers of special interest.

19 GEOPHYSICS, SEISMOLOGY

Books—Monographs—Reports

23596 **Canada, Department of Energy, Mines and Resources, Earth Physics Branch.** Canadian seismograph operations, 1980—Annuaire seismographique du Canada, 1980: Seismological Series (Ottawa)=Serie Seismologique (Ottawa), No. 86, 95 p., illus. (incl. 6 tables, sketch map), 8 ref., 1981.

23597 **Everingham, I. B.; and Sheard, S. N.** Seismicity of the New Guinea/Solomon Islands region, 1972: Report - Australia, Bureau of Mineral Resources, Geology and Geophysics, No. 220, 10 p., illus. (incl. 7 tables), 1980. (BMR Microform MF110).

23598 **Evernden, Jack F.** (U. S. Geol. Surv., Menlo Park, CA, United States). Summaries of technical reports, Vol. IV: variously paginated, July 1977. *available from:* U. S. Geol. Surv., Menlo Park, CA, United States. Submitted to National Earthquake Hazards Reduction Program; individual papers within scope are cited separately.

22 ENGINEERING AND ENVIRONMENTAL GEOLOGY

Individual Papers and Chapters

36544 **Whitman, Robert V.** (Mass. Inst. Technol., Cambridge, MA, United States). The eternal triangle; seismologists/geologists, engineers/architects and owners/ regulators: *in* Proceedings of Earthquakes and earthquake engineering; the eastern United States; two volumes (Beavers, James E., chairperson; *et al.*), Ann Arbor Sci. Publ., Ann Arbor, MI, United States, p. 1177-1189, illus., 1981. *Meeting:* Assessing the hazard; evaluating the risk, Sept. 14-16, 1981, Knoxville, TN, United States.

36545 **Whitman, Robert V.; Donovan, Neville C.; Bolt, Bruce; Algermissen, S. T.; and Sharpe, Roland L.** Seismic design regionalization maps for the United States: *in* Earthquake engineering research at Berkeley, 1976 (University of California at Berkeley, College of Engineering, Earthquake Engineering Center), Report - Earthquake Engineering Research Center, College of Engineering, University of California, Berkeley, California, No. 77/11, p. 19-22, illus., 8 ref., May 1977. *Meeting:* Sixth world conference on earthquake engineering, Jan. 10-14, 1977, New Delhi, India.

36546 **Whitman, Robert V.** (Mass. Inst. Technol., Cambridge, MA, United States); **Lambe, P. C.; and Kutter, B. L.** Initial results from a stacked ring apparatus for simulation of soil profile: *in* Proceedings of the International conference on recent advances in geotechnical earthquake engineering and soil dynamics; Vol. III (Prakash, Shamsher, chairperson), Univ. Mo., Rolla, MO, United States, p. 1105-1110, illus. (incl. 1 table), 6 ref., January

3, 1982. *Meeting:* Apr. 26-May 3, 1981, St. Louis, MO, United States.

18 GEOPHYSICS, SOLID-EARTH

Individual Papers and Chapters

23389 **Dehlinger, Peter** (Oreg. State Univ., Dep. Oceanogr., Corvallis, OR, United States); **and Couch, R. W.** Gravity investigations west of the coasts of Oregon, Washington, and northern California: *in* Upper Mantle Project; United States Program; final report (U. S., National Academy of Science, Upper Mantle Committee), p. 232-233, July 1971. *available from:* Natl. Acad. Sci., Upper Mantle Comm., Washington, DC, United States.

23390 **Dergunov, A. B.** Stroyeniye kaledonid i razvitiye zemnoy kory v Zapadnoy Mongolii i Altaye-Sayanskoy oblasti [Structure of the Caledonides and development of the Earth's crust in western Mongolia and the Altai-Sayan area]: *in* Problemy tektoniki zemnoy kory (Peyve, A. V., editor; *et al*), Izd. Nauka, Moscow, USSR, p. 183-193, strat. cols., geol. sketch map, 29 ref., 1981.

23391 **Dickinson, William R.** (Univ. Ariz., Lab. Geotectonics, Tucson, AZ, United States). Plate tectonic evolution of the southern Cordillera: *in* Relations of tectonics to ore deposits in the southern Cordillera (Dickinson, William R., editor; *et al*), Arizona Geological Society Digest, Vol. 14, p. 113-135, illus. (incl. sketch maps), 140 ref., 1981. *Meeting:* March 19-20, 1981, Tucson, AZ, United States.

23392 **Dohr, G.** (Preuss. AG, Erdoel und Erdgas, Hannover, Federal Republic of Germany). Deep crustal reflections and the structure of the Earth's crust: *in* Mobile Earth; International Geodynamics Project; final report of the Federal Republic of Germany (Closs, H., editor; *et al*), Harold Boldt Verlag, Boppard, Federal Republic of Germany, p. 155-160, illus., 8 ref., 1980.

Subject Index

The Subject Index, which nearly fills the second half of an issue, can lead you to references on general or specific topics. The terms in the Subject Index are general and specific concepts chosen to describe the content of the documents cited in the Bibliography. On the average you will find 3 to 4 entries in the Subject Index for each bibliographic citation. As you become more familiar with our indexing vocabulary (see Chapter 5, Alphabetical Term List) and with the indexing system, you will become a more effective literature searcher. The indexing system is discussed later in this chapter and in Chapter 4. The examples given here, however, introduce you to using the Subject Index.

Scanning for Specific Research Topics

You want to keep up to date on a specific subject, geographic area, fossil group, process. Find the topic in the Subject Index. Note the reference numbers following the key words, and use the reference numbers to find the citations in the Fields of Interest.

Topic: Sedimentation in the Baltic Sea
Subject Index:

Baltic Sea—oceanography
sediments: Suspended matter of the Baltic Sea Basin; 3, The distribution of suspended matter in near-mouth regions of the Baltic Sea under the influence of river plumes 10922

Fields of Interest:
10922 **Pustel'nikov, O. S.; and Yakubovich, V. V.** Vzveshennoye veshchestvo basseyna Baltiyskogo morya (3, Raspredeleniye vzvesi v priust'yevykh uchastkakh Baltiyskogo morya pod vliyaniyem rechnykh vynosov) [Suspended matter of the Baltic Sea Basin; 3, The distribution of suspended matter in near-mouth regions of the Baltic Sea under the influence of river plumes]: Liet. TSR Mokslu Akad. Darb., Ser. B: Chem., Tech., Fizine Geogr., No. 3 (118), p. 123-129 (incl. Lithuanian sum.), illus., 10 ref., 1980.

Topic: Jurassic Cyclostomata
Subject Index :

Bryozoa—Cyclostomata
Jurassic: Two new Jurassic bryozoa from southern England 11159

Fields of Interest:
11159 **Taylor, Paul D.** Two new Jurassic bryozoa from southern England: Palaeontology, Vol. 23, Part 3, p. 699-706, illus. (incl. tables, plates), 22 ref., August 1980. *Reptomultisparsa tumida, Reptoclausa porcata.*

Topic: Mineral resources in Colorado
Subject Index :

Colorado—economic geology
mineral resources: Energy and mineral resources of Colorado; their potential for the future 51163

Fields of Interest :
51163 **Murray, D. K.** (Colo. Sch. Mines Res. Inst., Golden, CO, United States). Energy and mineral resources of Colorado; their potential for the future: Colorado Communicator, Vol. 10, No. 6, p. 2-6, 30 ref., December 1980.

Author Index

In the Author Index, which concludes each monthly issue, names and initials of all authors of documents cited in that issue are listed alphabetically. Immediately to the right of each name, a reference number is given for the citation.

Keeping Up With Your Colleagues

If you know the names of geoscientists working in your subject area, you can find documents they have written either as senior author or as a co-author. Occasionally a document appears which only gives the name of the organization doing the work. In this case the organization is given as the author in the citation. This practice is called "corporate authorship" and is used when no personal author is given.

Searching Procedure

1. Locate author name in alphabetical position.
2. Using reference numbers, check citations in Fields of Interest.

7

Author Index

Grishina, L A 26019
Grocott, J 23125
Gromme, S 23419 23425
   23426

Fields of Interest

23419 **Gromme, Sherman** (U. S. Geol. Surv., United States); **and Hillhouse, John W.** Paleomagnetic evidence for northward movement of the Chugach Terrane, southern and southeastern Alaska: *in* The United States Geological Survey in Alaska; accomplishments during 1979 (Albert, N. R., editor; *et al.*), Geological Survey Circular, No. 0823-B, p. 70-72, sketch map, 8 ref., 1981.

Notes on the Author Index
●general:
● Alphabetization is word by word; thus, DEL MONTE appears before DELANEY.
● If a name differs from AGI's usage because of transliteration rules or other reasons, both forms will appear in the index.
● A separate entry is given for each author or co-author (no maximum).
●personal names:
● Family names are followed by initials of all first names given in the original document.
● Compound family names such as von Hippel, F., are listed under each component, e.g., von Hippel, F., and Hippel, F. von.
● Oriental names such as Xin Chen are also entered as Chen Xin.
● Family names with O', M', Mc, Mac have no space (O'CONNOR).
● Family hierarchy (Sr, Jr, I, II, III) notations follow the initials (BLACKSTONE, D L, JR).
● Hyphens and spaces are retained as in the original (ABDEL-AAL, O Y).
●corporate names:
● When a corporate body is the only responsible body source indicated in the source document it is used in the Author Index.
● Only well-known acronyms are used. These are listed with spaces between the letters (U N E S C O).
● Where several levels of an organization are cited they are entered in descending order of scale, from the larger unit to the smaller.
● If the corporate name is ambiguous, additional information is given (DEEP SEA DRILLING PROJECT, LEG 59 SHIPBOARD SCIENTIFIC PARTY).

SEARCHING THE ANNUAL CUMULATIONS
The *Bibliography and Index of Geology* and its forerunners contain more than a million citations. Finding the information you seek among so much data would be most difficult without the Subject Index the Bibliography provides. An understanding of the arrangement of material in the Bibliography, the indexing vocabulary, and a few guidelines for searching will help you.

How the Cumulative Volumes Differ from the Monthly Issues
In making a search you will move back through time, year by year, in the cumulative volumes. Each set (with the exception of Volume 41, 1977) consists of a 2-volume Bibliography and a 2-volume Subject Index. Each year's set is hardbound in a different color.
At the beginning of Part 1 (except 1977), you will find a complete list of Serials included in that year's Bibliography. If you are looking for a particular serial or, from 1980 on, for particular issues of a serial, you can scan the list to see whether they are included.
The annual Bibliography, parts 1 and 2, is an alphabetical listing by author of all citations included in the monthly issues. Unlike the monthlies, however, there is no grouping by document type or Fields of Interest. You will note also that the citation identifier number is not included in the annual volumes (except for 1977). In 1977, no cumulative bibliography was produced, and citation numbers given in the cumulative index guide searchers to the monthly issues.
The cumulative index, parts 3 and 4, is a massive compilation of the monthly subject indexes. The format of the entry remains the same. With a quick look at the Index you will see that entries for an important topic such as *ground water* fill many pages. The following guidelines should help you avoid getting bogged down in your search.
Guidelines for Searching
1. Make a clear statement of your research problem.
2. Select the most appropriate place to search: the Serials section, Bibliography, or Subject Index.
3. Devise a search strategy including terms to be searched and years to be searched with the help of this Guide.
4. Perform the search. Keep in mind that the Bibliography has changed over the years; changes not described in this chapter are covered in Chapter 4.
5. Modify your search strategy as you search, according to your results.
Search by Subject
Except for the current year, a subject search in the Bibliography is done using the annual cumulations. Entries in the Index are arranged by level-1 term, and under these by level-2 term, and under these by level-3 term.
Because our indexing uses a controlled vocabulary and 3-level index sets, to search efficiently it is important that you find the precise terms and sets you need. Index sets have the form:
Level 1—level 2
    level 3: Title, citation number
Special Term Lists
As you look at your research topic, you will probably find that it includes one or more key

concepts that provide starting points for your search.

If you seek information on any of the following topics, check the appropriate Special Term List(s) for a structured approach to choosing your search terms. These lists represent a systematic approach to important basic topics in geology. All terms in these lists are also in the Alphabetical Term List.

List A Level-1 terms
List B Geologic disciplines and area sets
List C Commodities
List D Elements
List E Geologic age (stratigraphic) terms
List F Fossils
List G General terms
List H Igneous rocks
List I Sedimentary rocks
List J Metamorphic rocks
List K Sedimentary structures
List L Minerals
List M Soils
List N Sediments
List O Geographic terms

● Are you looking for information on a geographic area? (See lists O and B)
● Does your topic cover a geologic age, stratigraphic feature, rock, mineral, or fossil? (See appropriate Special Lists)
● Are you seeking information on petroleum, coal, or another commodity? (See List C)

If you have found terms for your topic in the Special Term Lists then look them up in the Alphabetical Term List for useful information on set structure, dates of usage, and related terms. And if you haven't found an approach to your topic through the Special Term Lists, then go to the Alphabetical Term List and look there for the concepts in your topic.

Alphabetical Term List

The Alphabetical Term List contains all entry points in the Index, i.e., first-level terms and cross-references. In addition, it includes second-level terms. By checking the terms you intend to search in this list, you will get useful information on set structure, dates of usage, and related terms.

In using the Alphabetical Term List, consider which concept in the research topic is most likely to be found in a geology index. For example, if the topic is the use of remote transducer instruments for earthquake detection, among the concepts "remote transducer instruments", "earthquake" and "detection" choose "earthquake". The Alphabetical Term List under "earthquakes" lists "detection" as one of the second-level terms. This indicates that the place to look in the Index is under earthquakes—detection.

If you do not find the term you seek in the Alphabetical Term List, it may well be that the term is too specific. Try a broader term or a related term. Examples:

Specific Term Not in List	Broader or Related Term in List
Silver Bow County	Montana (List O)
hematite	iron ores (List C)
strip mining	mining
Mt. St. Helens	volcanoes and Washington
leaching	weathering and soils
Rotaliina	foraminifera (List F)
shear strength	soil mechanics, rock mechanics, and engineering geology
montmorillonite	sheet silicates (List L)

Broader terms of the geographic and systematic terms above can best be located in the Special Term Lists. Those unfamiliar with the systematic terminology of geology might look in a reference such as AGI's *Glossary of Geology* to determine the appropriate class of a specific fossil, rock, mineral, etc. And, of course, a gazetteer might be used for particular geographic terms. Many of these specific terms can also be found in the *GeoRef Thesaurus and Guide to Indexing*. This is the tool used by our indexers, from which the terms in the Alphabetical Term List are extracted. The third edition, 1981, contains approximately 13,500 terms and is available from Customer Services at AGI.

Many of the cross-references in the Alphabetical Term List are underscored to indicate that any index term can appear in that spot in the cross-reference. In making the cross-reference, the computer scans *all* index sets looking for candidate terms in specified relationships with other terms.

Index Modifications

Differences to be aware of in the Subject Index from 1969 on are described below:

● *citation locator*—To enable you to go from an index entry to the citation, in every year but 1977, the primary author is given. In 1977, a citation number is given instead, and the user must go to the monthly issues to find the citation. Citations are numbered from 1 to 48220. In all other years the citations were cumulated and sorted by author.

● *cross-references*—In 1978 cross-references began to be used in the Subject Index. These provided an approach to level-2 and level-3 terms, and to synonyms and related terms. Hence you might do well to begin with the most recent cumulation and work back. But in doing so, be aware that prior to cross-references some redundant sets were used which were eliminated with the advent of cross-references. In going back from 1978 to 1977, if you have been using a cross-reference from a term, e.g.,

Iron
   see also under economic geology
     under Africa; Angola; Austria; etc.

then instead look under the term itself for the

9

same information, e.g.,
Iron—Africa
     Egypt . . .
● *index sets*—In 1978, level 3 of the sets changed in that the string of terms used 1969 to 1977 was replaced by a single term followed by the document title. This title also supplements the author as a citation locator. The rules now in effect for structuring specific index sets are given under the level-1 terms in the Alphabetical Term List and in the Special Lists. Changes since 1969 in important classes of sets are given in the chapter "Subject Indexing".

A Sample Subject Search
You need as much information as possible on the depositional environment of the Mowry Shale. You already know that it is in the Colorado Group, is of Cretaceous age and is found in Montana, South Dakota, and Wyoming.

Since area terms are significant, go to List O—Geographical Terms. "Montana", "South Dakota" and "Wyoming" appear here as valid index terms. Next go to List B—Area Sets. Here you find that *stratigraphy* and *sedimentary petrology* are the second-level terms to look for under your area terms, Montana, South Dakota and Wyoming. The third-level term under *stratigraphy* will be an age term. In List E—Geologic Ages you find that Cretaceous is a valid term. Going back to List B you find that the third-level terms under *sedimentary petrology* that apply are *sedimentary rocks*, *sedimentation*, and *sedimentary structures*.
These would be sets having:
     Level 1: Montana *or* South Dakota *or* Wyoming
     Level 2: stratigraphy
     Level 3: Cretaceous

Montana—stratigraphy
Cretaceous: Extinctions; iridium and who went when 22500
— Upper Cretaceous-Paleocene biostratigraphy and magnetostratigraphy, Hell Creek and Tullock formations, northeastern Montana 22407

OR
     Level 1: Montana *or* South Dakota *or* Wyoming
     Level 2: sedimentary petrology
     Level 3: sedimentary rocks *or* sedimentation *or* sedimentary structures

sedimentary structures: Paleoenvironment of the Mowry Shale (Lower Cretaceous), western and central Wyoming, as determined from biogenic structures 06435

Indexing and Indexers
Indexing is a subjective process of choosing the best words from a controlled vocabulary to describe a paper, and then arranging them to provide the searcher several pathways to finding the reference. Although Chapter 4 covers our indexing system in detail, including changes in indexing practice from 1969 to the present, some background on the way an indexer handles a paper should help you formulate your searches.

The indexer begins by recording the bibliographic information on a color-coded worksheet and assigning the paper to one or more subject categories. As the indexer carefully notes the title, abstract, illustrations, maps, and type of document, these elements suggest appropriate key words. If there is no abstract, the indexer uses the introduction and conclusion of the paper to tell what is covered.

Looking at a paper, the indexer asks the same questions you should ask yourself about your research topic:

● Is it general or specific?
● Is it tied to a geographic area and/or age?
● Is it about a rock, fossil, commodity, or other topic covered in the Special Lists?

Indexers say that their interpretation of a paper and the index terms they choose to describe it reflect the bias of their geological studies and interests. The controlled vocabulary, structured indexing system, and careful work of the indexers help minimize the variations and provide consistent, accurate indexing of the geoscience literature. The indexing is then checked by the Chief Editor who adds consistency to it, and who holds weekly meetings with the indexers to discuss indexing practices. In the end, the indexer's decisions are influenced by knowledge of the subject, experience, and awareness of current indexing practices. Indexers are assigned various subject areas and journals to index regularly on the basis of their geological backgrounds. The indexers have degrees in geology and work full time at AGI where their primary responsibility is indexing.

The indexer has the option of using as many as 10 index sets to describe the content of a paper. In general, such a large number will only be used for a paper that covers several subject areas in depth. An example might be a paper that discusses petrology, structure, continental margins, ophiolites, and geochemistry. Or, if a paper is so poorly written that its key topics are not obvious, the indexer will make extra sets so that nothing of importance will be overlooked. The following example, by contrast, is straightforward:

The indexer is holding a high-resolution aeromagnetic map of Quebec. After recording the bibliographic data, the indexer makes 2 sets, knowing that automatic cross-referencing will create 3 additional entries in the Index.

Set 1:
Quebec
  geophysical surveys
    maps

This set will appear in the Index and will generate the following cross-references:
maps
  see under geophysical surveys
    under Quebec

Set 2:
Quebec
  geophysical surveys
    magnetic surveys

This set will also appear in the Index and will generate the following cross-references:
magnetic surveys
  see under geophysical surveys
    under Quebec

geophysical surveys
  see magnetic surveys
    under geophysical surveys
    under Quebec

Thus, a researcher could find this map from 4 starting points in the Index: Quebec, geophysical surveys, maps, and magnetic surveys.

The indexers refer to an expanded version of the Alphabetical Term List, the *GeoRef Thesaurus and Guide to Indexing*, that includes third-level terms, and to the Special Term Lists in Chapter 6, to determine which form of a term to use and how to use it.

Search by Author

If you plan a retrospective search by author, you should be aware of changes in the handling of author names in the cumulations. There is no multiyear cumulated author index.

● *alphabetization*—The authors are alphabetized by surname, and within surname by given names. Within both surnames and given names, punctuation is ignored but spaces are not. This produces word-by-word alphabetization within surname and within given names. Consequently, for example, Smith, A. Z. appears before Smith, Anthony, because the space following A. sorts before An.

● *anonymous*—Most references in the Bibliography have a personal author. Those which do not will appear under "Anonymous" or under a corporate author (see below).

● *Chinese names*—Prior to 1981 the several parts of Chinese names were not cross-referenced. Since 1981 they have been.

● *compound surnames*—From 1969 to 1980, compound surnames appeared under the part of the surname where the usage of the country of

the author would put it. For example, an American named J. De Vries would appear under "De Vries, J." but a Dutchman with the same name would appear under "Vries, J. de" (1969–1974) or under "de Vries, J." but sorted under Vries (1975–1980). No cross-reference was provided from the unused part of the surname to the used. From 1981 on, the citation carries the uninverted form of the surname (e.g., de Vries) and is sorted as "de Vries" and there is a cross-reference from Vries, de, to "de Vries". However, if the names are hyphenated, no cross-reference is made from the part which follows the hyphen. See also Chinese names.

● *corporate authors*—"Canada, Geological Survey", "U. S. Geological Survey", and similar entries are used. Fairly often in the past "anonymous" was chosen where a corporate author might have been used, but recently there has been increased use of corporate authors. In format the largest unit of an organization is followed by the smallest unit, with or without intermediate units in between.

● *diacriticals*—From 1969 to 1971 diacriticals were used, e.g., for the umlaut in Müller. These were ignored in alphabetization. Beginning in 1983, diacriticals will reappear in the Bibliography.

● *joint authors*—From 1969 to 1981, only the first 3 authors of a paper appeared in the reference. If there were more than 3 this was indicated by et al., but cross-references were made from all the joint authors to the reference. From 1982 on, all authors' names appeared in the reference and the cross-referencing from each continued as before.

● *regularization of the authors' names*—From 1969 to 1975 the full name of the author was used if given in the document. An attempt was made in each cumulation to regularize authors names. This tended to the use of initials, when in doubt; e.g., if Eugster, H. P., occurred in one paper and Eugster, Hans P., in another, and they seemed from the context to be the same person, the name would be regularized to Eugster, H. P. From 1976 to 1980, only initials were used. From 1981 on, the name was entered exactly as in the source document and no attempt at regularization was made.

The Library—The Last Step in Searching and Scanning

As you work, you will modify your search according to the references you find. If you need to see certain original documents, your library is the best source. Using the bibliographic details we have provided, the librarian may suggest several methods of obtaining the documents. These may include use of your library's periodical collection, borrowing from another library, or using a document-delivery service which provides originals as copies or film for a service fee.

2. DOCUMENTS COVERED

Many types of documents are cited in the Bibliography. This chapter lists the types and for each describes the extent of coverage and other factors you need in order to evaluate the coverage in the light of your requirements. Documents and features not covered in GeoRef are also listed.

General Criteria for Inclusion
—Original research in geology, including experimental and theoretical research and field studies
—History, philosophy, and practice of geology
—Subject reviews and current research reports
—Geological reference materials

Specific Sources Covered
The following types of documents as they relate to geology are included:

1. Serials
2. Monographs
3. Maps
4. Theses
5. Meeting papers
6. Reports
7. Textbooks
8. Biographies
9. Bibliographies
10. Guidebooks
11. Catalogs
12. Museum publications
13. Association publications
14. Reprints
15. Translations
16. Annual reports

These are discussed in turn below. It should be noted that we categorize a document either as a serial or a monograph and the document may fit one or more of the other types as well.

1. Serials
Serials include periodicals, newspapers, journals, annuals (reports, yearbooks, directories, etc.), recurring conference proceedings or society transactions, and monographic series.

A serial is a document in print or non-print form, issued in successive parts, usually having numerical or chronological designations, and intended to be continued indefinitely. This is based on the definition of the International Center for the Registration of Serials in *Guidelines for ISDS* [International Serials Data System] (Paris, Unesco, 1973. (SC/WS/53()).

It should be noted that this definition covers unnumbered series but excludes collections which are not intended to be continued indefinitely.

A cumulated list of serials cited in GeoRef, from 1967 to date, is prepared annually. This is the *GeoRef Serials List and KWOC Index*, available from Customer Services, American Geological Institute. The June 1982 edition is $80 paper (unbound, prepunched, 1497 p.) and $15 microfiche (24x, 31 fiche).

The List is arranged by full title and includes for each serial complete title or ISDS Key Title, abbreviated title or imprint, CODEN and/or ISSN (if available), country of publication, and, for some, notes on title changes, etc.

The accompanying KWOC index permits lookup by each significant title word. It thus gives a positive indication whether or not a serial was covered, independent of the first words in the title.

The KWOC Index is also a tool for finding geoscience serials by subject and organization. For example, over 300 serials are listed therein which contain a word beginning with "hydro-", "water-", or "wasser-"; over 250 with "geophy-", "geofis-", or "geofiz-"; 24 beginning with "volcan-" or "vulcan-"; and over 80 with "Poland", "Polish", "Polon-" or "Polsk-". Under the term "Survey", publications of geological surveys from many countries are listed.

In addition to the cumulated Serials List, each monthly issue and cumulation of the Bibliography contains its own Serials section consisting of all serials cited in that issue or cumulation.

For each serial scanned, articles within the scope of the Bibliography, as reflected in the Fields of Interest, are selected. With some serials, this means virtually every article. For others, such as general journals like *Science* or journals in peripheral fields such as *Analytical Chemistry*, only a few articles are selected.

It is our policy to scan a wide range of serials. The current *Serials List and KWOC Index* lists over 7,700 serials. The list is open-ended. New serials with articles in scope are regularly added.

In the 1980 annual cumulation of the Bibliography, 2,432 different serials were cited. Other serials were scanned in 1980 but no articles from them were selected.

Information on which volumes of a specific serial were included in GeoRef is available in the GeoRef Library. The Library also retains information on the full title, publisher, ISSN, CODEN, history and frequency of serials in GeoRef, although most of this information is also in the GeoRef Serials List. Feel free to call the librarian for this information.

A few serials, mostly publications of AGI member societies, are cited quickly from prepublication proofs sent to GeoRef for this purpose. Other serials given priority are those cited in *Minerals Exploration Alert* and those for which annual indexes are produced from GeoRef.

If there is a question as to whether or not a

given publication is a serial, our inclination has been to treat it as a serial. This is the case for proceedings of certain international meetings held in a different city each year, and for series of maps published by geological surveys, for open-file reports and for series of government reports. An exception is field-trip guidebooks, which are usually cited as monographs.

2. Monographs

These include books and other publications which are not parts of series. They are cited as individual pieces, and, if they contain separately authored chapters, the chapters are cited as well. A note in the monographic citation often indicates whether or not the chapters are also cited. For example:

> 11552 Fanning, Kent A. (Univ. South Fla., Tampa, FL, United States); and Manheim, Frank T. (editors). The dynamic environment of the oceanfloor: D. C. Heath and Co., Lexington, MA, United States, 482 p., illus. (incl. 38 tables, sketch maps), 7(9 ref., 1982. Individual papers within scope are cited separately.

In addition to the above citation, 23 citations of papers in the book appear in the March 1982 Bibliography.

Our principal source of monographs has been the library at the National Center of the U.S. Geological Survey, which is now located in Reston, Va. This library has the most comprehensive collection of geological literature in the U.S. All monographs acquired by this library are considered for inclusion in GeoRef and the Bibliography, based on the criteria of scope reflected in the Fields of Interest. The policy has been, and continues to be, to cite any monographs within scope.

The U.S. Geological Survey Library is, naturally, strongest in North American monographs, but its coverage of other parts of the world is quite good as well. Beginning with those published in 1981, monographs have been complemented by inclusion in GeoRef and the Bibliography citations received in our exchange program with the Europeans through BRGM/CNRS.

Since 1978, monographs are grouped in the monthly issues of the Bibliography under the subheading "Books—Monographs—Reports" within each Field of Interest.

From 1967-1974 the annual cumulations of the *Bibliography* included a list called Special Publications, which contained books and other monographs covered for the year. This list begins to appear again in the 1983 annual cumulation.

3. Maps

Both separately published maps and maps which are part of other publications are included in the Bibliography.

Separately published maps—The emphasis has been on maps of the U.S. and Canada, although maps of other areas are also well represented. From 1981 on, coverage of maps outside of North America, and particularly of maps of Europe, has been complemented by the exchange of references with the Europeans through BRGM and CNRS.

Maps produced by the U.S. and Canadian geological surveys are thoroughly covered, as are those of the state surveys. But the coverage extends to geological surveys and other publishers throughout the world. Separately published maps have their own citations in the Bibliography. An example from the March 1982 Bibliography:

> 12315 Ortman, B. H. (Geol. Surv. Can., Canada). Geology; Bell River, Yukon Territory—Northwest Territories: Map-Geological Survey of Canada, No. 151)A, 1 sheet, colored geol. map 1:250,000, 1981.

In the monthly issues, maps appear under the subheading "Separately Published Maps" in each Field of Interest, and particularly in Field 14, "Areal Geology, Maps and Charts". Also, separately published and some of the more significant maps which appear in other publications can be found in the Subject Index under "maps" where they are subdivided by or cross-referenced to areas.

Maps which are parts of other publications—Many geological publications contain maps. The inclusion of maps, the kind of maps, and the scale, if given, are indicated in the references in the Bibliography. Often the phrase 'sketch maps' is used in citations to indicate less elaborate maps usually smaller than a page. Larger maps are particularly likely to be found in monographs. For example, the following citation appears in Field 24, Surficial Geology, Quaternary Geology, under the subheading "Books—Monographs—Reports" of the March 1982 Bibliography:

> 14300 van Staaldvinen, C. J. (editor) (Rijks Geol. Dienst, Haarlem, Netherlands). Geologisch onderzoek van het Nederlandse Waddengebied [Geologic investigation of the Netherlands' Wadden region]: Rijks Geol. Dienst, Haarlem, Netherlands, 77 p., illus. (incl. sect.; Quaternary geol. maps; colored geol. map 1:50,000) 1977. Individual papers are cited separately.

4. Theses

Carrying on a task performed for years by Dederick Ward for the Geoscience Information Society, each year we send a letter to geology departments in the U.S. and Canada requesting abstracts of theses for degrees awarded that year in geology and related sciences. The request is

13

for doctoral dissertations (if unreported to Dissertation Abstracts), masters' theses, and, if any, bachelors' theses. The response to these letters is good. All theses reported are added to GeoRef and cited in the Bibliography. In addition, the geology and geology-related sections of Dissertation Abstracts are scanned and citations added for all dissertations which are in scope. Few theses from schools in countries other than the U.S. and Canada have been covered in the Bibliography thus far. However, in the future, theses from European schools will be included through our exchange of citations with the Europeans through BRGM and CNRS.

Since 1978, theses can be found in the monthly issues under the subheading "Dissertations and Theses".

5. Meetings

Publications which come out of meetings are included. Our preference is to consider meeting proceedings, symposia, transactions, etc., as serials. But some meetings are one-time affairs, and these we treat as monographs. In the *GeoRef Serials List and KWOC Index*, many of the serials are meeting series.

If the papers presented at meetings are gathered and published, each paper within scope is cited in GeoRef and in the Bibliography. However, single reprints of meeting papers are usually not cited.

When the meeting papers are published as a separate volume, a citation is made for the volume itself as well as for the individual papers. This is not done for individual meeting papers which appear as part of a publication that contains other papers, etc.

For some meetings a book of abstracts appears. From 1969 through 1976, abstracts of meeting papers were cited in the Bibliography whether or not the full papers were to be published as conference proceedings. If full papers later appeared, they were cited also. After 1976, for conferences such as the Lunar Science, for which the full papers were published as conference proceedings, the abstracts were not cited but the full papers were. International Geological Congress abstracts have been covered in their entirety as have the Geological Society of America's Abstracts with Programs. Foreign-language abstracts have not been covered since 1978 except in the case of international meetings.

Since 1978, specific meetings are listed in the monthly issues of the Bibliography under the subheading "Meetings" in each Field of Interest, and citations of meeting papers include meeting information. They also appear in the Subject Index under the term "symposia", where they are arranged by topic.

6. Reports

A report is an item, usually not available through normal commercial channels, but obtainable from the organization responsible for its issue. It is usually identified by a report number and usually a government document.

Among the reports covered are those of the U.S. Geological Survey, the 49 state surveys, and the Bureau of Mines. The Department of Energy reports from Grand Junction, Colo., are also included. Open-file reports of the U.S. Geological Survey are covered from 1978 on. Bureau of Mines open-file reports, which were not covered through 1981, were first covered selectively in 1982.

In addition to the above, Canadian Geological Survey and provincial publications are included, but the coverage is not complete. Coverage of publications of other governments is spotty. Coverage of European government publications should improve for 1981 and subsequent years, based on our exchange with the Europeans through BRGM and CNRS.

7. Textbooks

Secondary-school through graduate-school textbooks are cited. These include books sent to *Geotimes* for review and books purchased for the USGS library in Reston, Va. These appear in the Bibliography Subject Index under the level-2 heading "textbooks" under geological discipline terms such as "mineralogy" and "seismology".

8. Biographies

A substantial number of biographies and obituaries of geologists are cited each year in the Bibliography. However, inclusion is selective. Practice is to cite only those biographies which include substantial bibliographies. Biographies are found under the heading "Biography" in the Subject Index.

9. Bibliographies

Most bibliographies cited in GeoRef are monographs, but some which appear in serials are also included. These can be found in the Subject Index under the term "Bibliography", where they appear subdivided by discipline. Other bibliographies of papers on specific geographical locations appear under the area terms in the Index. From 1977 on, these area bibliographies can also be found under the term "Bibliography" as cross-references. Since 1979, the number of references in the document cited is given as part of each citation in the Bibliography.

10. Guidebooks

Geological field-trip guidebooks and road logs are included in GeoRef. Some are treated as serials, some as monographs. Guidebooks are difficult to keep up with. Many are published by the organizers of the field trips or the meeting sponsors, and never get into the ordinary publication channels. We have relied mainly on the USGS Library, Reston, for these. The library has a good collection. In addition, we are beginning to include guidebooks acquired by the libraries that contribute to the *Union List of Field-Trip Guidebooks of North America*, which is published by AGI from data supplied by the Geoscience Information Society.

Guidebooks for an area can be found in the Bibliography by looking in the Subject Index under the appropriate level-1 area term, then under "areal geology" on level 2, then under "guidebook" on level 3. For example:

Alberta—areal geology
guidebook: CSPG international conference; Facts and principles of world oil occurrence; Geological guide to the central foothills and Rocky Mountains of Alberta (Jones, P. B., et al.)

11. Catalogs

These appear in the Subject Index under the term "catalogs", arranged by discipline, for example:

catalogs—paleontology
foraminifera: Illustrated Key to genera of Paleozoic smaller foraminifera of the United States (Conkin, J. E., et al.)

Catalogs that are published annually, monthly, weekly, etc., and do not have separately authored papers, have not been cited since 1978.

A second place to look for catalogs is under "museums" (see below).

12. Museum Publications

Books, etc., about museums are found in the Subject Index under "museums". Catalogs of museum collections and exhibits are included here.

13. Association Publications

Under "associations" in the Subject Index are found descriptions and reports of current research, etc., issued by associations. The term "associations" is used to include foundations, institutes, societies, academic institutions and government agencies that produce reports on subjects within the scope of the Bibliography.

14. Reprints

Reprint volumes (not reprints of individual meeting papers) were routinely cited prior to 1977. Since then, reprints are included only when the original documents have not been cited.

15. Translations

In general, English translations from any language are included, as are the original documents from which they are translated. However, from 1976 on for journals which have cover-to-cover translations in English, only the English translations are cited. Translations from English to other languages are not included.

16. Annual Reports

Presently these are included when they contain separately authored papers. Before 1977, they were usually cited whether or not they contained individually authored papers.

Specific Exclusions

The following are not included in GeoRef or in the Bibliography:

—Errata, except when of significant length
—News items are routinely excluded. However, beginning in 1982 an exception has been made and we have begun to include some news items on minerals economics and minerals exploration.
—Book reviews
—Newsletters have been excluded since 1977 unless they contain separately authored papers.
—Letters, unless they contain substantial information.
—Title-only sources—If only a document title is available to our indexers, the document is not cited. We either work from the source documents, or, for some meetings and for theses, from abstracts. An exception is made for citations of old, inaccessible documents for retrospective bibliographies, e.g., the Colorado 1875–1975 bibliography, for which some citations are done from titles only.
—Regional Russian geology—Between 1977 and 1981, journals published by regional scientific societies in the USSR were excluded unless needed for the Bibliography and Index of Micropaleontology or the Bibliography of Fossil Vertebrates. Also, since 1977, chapters of Russian monographs dealing with regional geology have been excluded. The books are cited only as monographs.
—Editorials
—Photographs without accompanying text
—Patents

3. FIELDS OF INTEREST

All references selected for inclusion in the Bibliography and Index of Geology must be assigned to a specific field of interest. In the monthly issues, references are arranged under these fields to allow the user to scan sections of particular interest.

CURRENT FIELDS OF INTEREST
There are currently 29 fields of interest. It is important to note that beginning in August 1981, references could be assigned to multiple fields of interest. When this occurs, a 'see-also' note appears at the end of the field.
The current list of 29 fields is an expansion of the 21 fields used in the Bibiography and Index of Geology from 1969 through 1974. Both current and former fields are listed in this chapter, followed by a chart relating the 21 to the 29 at the end of this section.
It is important to remember that the fields of interest are in the monthly issues of the Bibliography, but do not appear in the annual cumulations.
For easy reference, all topics listed below are included in the Index of this Guide.

01—Mineralogy and crystallography
physical, optical, and chemical properties of naturally occurring inorganic minerals, and related synthetic minerals
mineral crystallography, including crystal structure, determination of lattice parameters and unit cells, the bonding of atoms and molecules, crystal form and symmetry
collecting minerals, as well as non-mineral gems such as amber and jet
02—Geochemistry
abundance of elements
organic materials
water
trace elements
isotopes
geochemical processes and properties
geochemical cycles
geochemical surveys
analytical methods such as chemical, spectroscopic, thermal, and X-ray and electron microscopy
instruments used for analysis
Related topics in other fields:
Field 05—petrology, fluid inclusion, geologic thermometry and barometry, meteorites
Field 03—geochronology
Field 06—clay mineralogy
Field 12—paleoclimatology
Fields 26 & 28—geochemical prospecting
967 to 1974, some of the above topics

were included in Field 02, which covered a wider range of topics.)
03—Geochronology
determination of absolute age
methods, including radiometric and radiogenic dating
establishing the chronology of events by methods such as lichenometry, racemization, tree rings, hydration of glass, varves, paleomagnetism, tephrochronology, thermoluminescence, radiation damage, fission-track dating and particle-track dating
04—Extraterrestrial geology
of the planets, asteroids, the Moon, and moons of other planets
planetary composition, evolution, surface features, structure, motions, gravity, magnetic fields and atmosphere
Exclusions:
astrophysics and extraterrestrial physics
Related topics in other fields:
Field 05—meteorites and tektites (before 1980 these were in Field 04.)
05—Petrology, igneous and metamorphic
igneous rocks
metamorphic rocks
metasomatism
metamorphism
phase equilibria
magmas
lava
intrusions
inclusions
ancient volcanology
Related topics in other fields:
Field 23—volcanic features
Field 24—modern volcanology
06—Petrology, sedimentary
sedimentary rocks
sediments
sedimentation
diagenesis
sedimentary structures
genesis of peat, lignite and coal
clay mineralogy
chemical properties of clay minerals
Related topics in other fields:
Field 07—marine sedimentation
Field 24—Quaternary sediments
Field 29—economic studies of peat, lignite and coal
07—Marine geology and oceanography
ocean floors
ocean basins
ocean waves (sediment transport)
ocean circulation (sediment transport)
continental shelf

continental slope
Exclusions:
patterns of ocean circulation
marine biology
wave propagation
Related topics in other fields:
Field 02—geochemistry of sea water
Field 18—ocean basin evolution
Field 21—estuarine studies
Field 24—Quaternary sediments and recent
changes of level

08—Paleontology, general studies of both fossil plants and animals
life origin
paleontological textbooks
paleontological glossaries
fossil collecting
conodonts
problematic fossils
ichnofossils (if not related to a specific fossil group)
Related topics in other fields:
Field 09—fossil plants
Field 10—invertebrates
Field 11—vertebrates
Field 12—biostratigraphy

09—Paleontology, paleobotany
algae
angiosperms
bacteria
bryophytes
fungi
gymnosperms
lichens
palynomorphs
Plantae
pteridophytes
thallophytes
Related topics in other fields:
Field 12—biostratigraphy
Field 24—Quaternary palynology

10—Paleontology, invertebrate
Coelenterata
Echinodermata
foraminifera
Graptolithina
Hemichordata
Insecta
Mollusca
Ostracoda
Porifera
Pterobranchia
Radiolaria
Trilobita
worms
Related topics in other fields:
Field 12—biostratigraphy

11—Paleontology, vertebrate
Amphibia
Aves
Chordata
fossil man

Mammalia
Pisces
Reptilia
Related topics in other fields:
Field 12—Pre-Quaternary archaeology and artifacts
Field 24—Quaternary archaeology and artifacts

12—Stratigraphy, historical geology and paleoecology
lithostratigraphy (age relationships of rock strata)
biostratigraphy
evolution of land masses (continental drift)
paleomagnetism
paleogeography
biogeography
paleoclimatology
Related topics in other fields:
Field 06—reefs and sedimentation (From 1969 to 1974 was in Field 12)
Field 08–11—paleontology
Field 24—Quaternary stratigraphy

13—Areal geology, general
area studies dealing with several aspects of geology
entries that might be placed in three or more fields
guidebooks
road logs
bibliographies of geology of an area
Related topics in other fields:
Field 14—geologic maps (before 1975 these were in Field 13)

14—Areal geology, maps and charts
separately published geologic maps
geologic maps with explanatory texts
separately published geologic charts
methodology of geologic mapping
Related topics in other fields:
specific types of maps are found under the specific field, e.g., geomorphologic maps are found in Field 23, Surficial geology, geomorphology
Field 13—before 1975 maps and charts were in Field 13

15—Miscellaneous and mathematical geology
biography (if not related to a specific field)
bibliography (if not related to a specific field)
popular geology
elementary geology (textbooks)
general mathematical principles
annual reports of geologic surveys and associations
historical accounts education—curricula, enrollments, directories to geology departments
geology as a profession—career opportunities
forensic geology—geology applied to crime solving
Related topics in other fields:
anything related to a specific field would be in that field rather than here

17

16—Structural geology
classical tectonics (regional and local structures resulting from solid-rock movements)
faults, fractures, folds, orogeny, geosynclines, deformation, structural analysis, epeirogeny, foliation, lineation
Related topics in other fields:
Field 18—plate tectonics, continental drift, sea-floor spreading
Before 1975 the following subjects were found in Field 16; they are now found in other fields:
Field 05—batholiths, dikes, intrusions, stocks and volcanism
Field 06—breccia, sedimentary structures
Field 12—changes of level, paleogeography, unconformities
Field 18—crust, diapirs, geodesy, isostasy, Mohorovičić discontinuity
Field 22—nuclear explosions
Field 23—cratering, cryptoexplosion features
Field 24—glaciers
17—Geophysics, general
experimental and theoretical studies of physical properties of rocks transition states of various compounds under elevated temperature and pressure (applied to core and mantle composition)
magnetic and electrical properties of minerals and melts that relate to the Earth's magnetic field
history, development and education in geophysics (since 1981 in Field 18)
magnetic and gravity fields of the Earth (since 1981 in Field 18)
Related topics in other fields:
Field 04—Extraterrestrial geology
Exclusions:
Meteorology, magnetosphere, astrophysics, aeronomy and solar physics
18—Geophysics, solid-Earth
worldwide structure of the Earth
plate tectonics
continental drift
sea-floor spreading
paleomagnetism
structure of core, crust and mantle
Mohorovičić discontinuity
Related topics in other fields:
Field 16—Regional and local structure
Prior to 1975 the following subjects were found in Field 18; they are now found in other fields:
Field 20—Geophysical surveys
Field 19—Seismology
Field 17—Gravity and magnetic fields, Earth's orbit, rotation and internal processes
19—Geophysics, seismology
Earthquakes and elastic waves, including seismograms, wave velocity, seismic sources, seismicity, microearthquakes, microseisms, mechanism of tsunamis and volcanic earthquakes
Related topics in other fields:
Field 18—before to 1975 all seismology papers

were in Field 18
Field 22—geologic hazards, seismic risk
20—Geophysics, applied
Applied studies not related to a specific subject including: well-logging, remote sensing, magnetotelluric surveys, gravity surveys, electrical surveys, magnetic surveys, seismic surveys, electromagnetic surveys
Related topics in other fields:
Field 18—before 1975 all geophysics papers were in Field 18
21—Hydrogeology and hydrology
ground water—pollution, geochemistry, movement, resources, mathematical models
thermal waters
springs
geysers
fumaroles
surface water—chemistry, sediment transport, hydrologic cycle (from 1969 to 1974, fewer references to surface water were included)
Exclusions:
biology of surface water
hydraulics
Related topics in other fields:
Field 02—geochemistry of water
Field 22—ground-water pollution was included in Field 22 until 1981
22—Engineering and environmental geology
geologic hazards—earthquakes, floods, land subsidence, landslides, debris flows, tsunamis, volcanoes
conservation
land use
pollution of surface water—before 1981 ground- water pollution was also found here (see Field 21)
waste disposal, reclamation
structures—dams, foundations, highways, marine installations, nuclear facilities, reservoirs, tunnels, underground installations and waterways when geologic subjects such as rock and soil properties are discussed
With the growth of the environmental sciences, inclusion of environmental topics has increased dramatically in recent years.
23—Surficial geology, geomorphology
genesis and evolution of features on the Earth's surface—meteor craters, cryptoexplosion features, eolian features, erosion features, fluvial features, frost action, lacustrine features, mass movements, shore features, solution features, and volcanic features.
Related topics in other fields:
Field 24—Quaternary glacial features
24—Surficial geology, Quaternary geology
the last 2 million years of Earth's history, including:
glacial geology
stratigraphy
palynology
modern volcanology

Before 1975 these topics were found in fields 23, 12 and 06.

25—Surficial geology, soils
soils—genesis, morphology, evolution, water regimes, chemistry, erosion and classification
Exclusions:
agricultural studies
Related topics in other fields:
Field 22—soil pollution, engineering properties of soils
Field 23—erosion processes

26—Economic geology, general and mining geology
commodity studies—more than one type of commodity mining geology
Related topics in other fields:
Field 22—mining engineering
Field 27—metals
Field 28—nonmetals
Field 29—energy sources

27—Economic geology, metals
metal ores—genesis, resources, economics, exploration, production and utilization (includes uranium)
Related topics in other fields:
Field 26—before 1975 all commodity studies were in Field 26

28—Economic geology, nonmetals
nonmetal deposits—genesis, resources, economics, exploration, production, and utilization
Related topics in other fields:
Field 26—before 1975 all commodity studies were in Field 26

29—Economic geology, energy sources
petroleum
natural gas
coal (economic studies)
oil shale
geothermal energy
oil sands
Related topics in other fields:
Field 26—before 1975 all commodity studies were in Field 26
Field 06—genesis of coal, peat, and lignite

PRE-1975 FIELDS OF INTEREST
Between 1969 and 1974 there were 21 Fields of Interest as follows:

01—Areal Geology
Economic geology, engineering geology, extraterrestrial geology, geochemistry, geochronology, geomorphology, geophysics, hydrogeology, igneous and metamorphic petrology, mineralogy, oceanography, paleontology, paleobotany, sedimentary petrology, stratigraphy, structural geology.

02—Economic Geology
Commodities (see Appendix D for a list of specific terms), mineral deposits, genesis, mineral exploration.

03—Engineering Geology
Clays, dams, earthquakes, experimental studies, foundations, gas storage, highways, landslides, materials, permafrost, rock mechanics, shorelines, soils, subsidence, tunnels, waste disposal, environmental geology.

04—Extraterrestrial Geology
Asteroids, moon, meteorites, Mars, Venus (and other planets), tektites.

05—Geochemistry
Biogeochemical prospecting, chemical analysis, chemical elements, clay mineralogy, crust, crystal chemistry, diffusion, electron microscopy, fluid inclusions, geochemical prospecting, geochemical surveys, geochemistry, geochronology, geologic thermometry, ground water, hydrothermal alteration, isotopes, laterites, magmas, major-element analyses, metasomatism, meteorites, mineral exploration, organic materials, paleoclimatology, petrology, phase equilibria, radioactivity, sea water, soils, spectroscopy, tektites, trace-element analyses, weathering, X-ray-diffraction analysis.

06—Geochronology
Absolute age, correlation, isotopes, paleoclimatology, paleomagnetism, paleontology, tephrochronology, time scales.

07—Hydrogeology
Artesian waters, connate water, fumaroles, geysers, glaciers, ground water, infrared surveys, mud volcanoes, permeability, porosity, springs, thermal springs.

08—Geomorphology
Basins, caves, erosion, estuaries, glacial geology, glaciation, glaciers, lakes, maps, mud volcanoes, rivers, shorelines, soils, weathering, permafrost.

09—Igneous and Metamorphic Petrology
Batholiths, breccia, cratering, cryptoexplosion structures, dikes, fluid inclusions, fumaroles, geologic thermometry, geothermal energy, orogeny, paragenesis, petrofabrics, petrology, phase equilibria, boudinage, foliation, geologic thermometry, lineation, metamorphic rocks, metamorphism, metasomatism, mineral zoning, igneous rocks, inclusions, intrusions, lava, magmas, meteorites, stocks, tektites, veins, volcanism, volcanoes, weathering.

10—Oceanography
Continental shelf, continental slope, estuaries, genesis, geophysical methods, marine geology, ocean basins, ocean waves, oceans circulation, ocean floors, reefs, sea water, sedimentation, sediments, seismic surveys.

11—Mineralogy
Crystal structure, crystal growth, crystal chemistry, crystallography, minerals.

12—General
Bibliography, current research, education, maps, philosophy, symposia, textbooks, catalogs, biography, associations.

13, 14, 15, 16—Paleobotany, Paleontology
Artifacts, collections, ecology, evolution, micropaleontology, organic materials, paleobotany, pa-

19

leoecology, paleontology, palynology, reefs, tracks and trails. (See also Appendix D for names of geologic periods.)

17—Sedimentary Petrology
Breccia, concretions, diagenesis, heavy minerals, marine geology, nodules, type of material, sedimentary structures, sediments, sedimentary rocks, sedimentation.

18—Soils
Chemistry, engineering properties, laterites, organic materials, permeability, porosity, sedimentation, soil group, weathering.

19—Solid Earth Geophysics
Geophysical surveys, seismology, earthquakes, tectonophysics, deformation.

20—Stratigraphy
Biostratigraphy, changes of level, continental drift, ecology, geochronology, geosynclines, glaciation, ice ages (ancient), lithostratigraphy, orogeny, paleobotany, paleoclimatology, paleoecology, paleogeography, paleomagnetism, paleontology, palynology, reefs, sedimentation, stratigraphy, unconformities.

21—Structural Geology
Basins, batholiths, boudinage, breccia, changes of level, continental drift, cratering, crust, cryptoexplosion structures, deformation, diapirs, dikes, epeirogenesis, faults, folds, foliation, fractures, geodesy, geosynclines, glaciers, intrusions, isostasy, lineation, meteor craters, Moho-

rovičić discontinuity, nuclear explosions, orogeny, paleogeography, petrofabrics, salt tectonics, sedimentary structures, stocks, tectonics, unconformities, uplifts, volcanism.

MAPPING OF OLD FIELDS TO NEW

Pre–1975	1975 to Date
1	13, 14
2	26, 27, 28, 29
3	22
4	4
5	2
6	3
7	21
8	23, 24
9	5
10	7
11	1
12	15
13	9
14	8
15	10
16	11
17	6
18	25
19	17, 18, 19, 20
20	12
21	16

4. SUBJECT INDEXING

The indexing is based on a system originated by John M. Nickles of the U.S. Geological Survey for the *Bibliography of North American Geology*. The system is also used in the Geological Society of America's *Bibliography and Index of Geology Exclusive of North America*, in *Geophysical Abstracts*, and in the *Bibliography of Theses in Geology*.

The basic unit of the system is a 3-level index set, such as:

Colorado
   economic geology
      coal: Methane in Cretaceous and Paleocene coals of western Colorado

For a document specific to an area, an area set is required, such as the above. Then, depending on the geological field of the document, other sets are prescribed. Some of the more common of these are described below.

In 1978, there were 2 important changes in the Subject Index. The string of index terms on level 3 was replaced by a single term plus the title of the document. And cross-references were added to related level-1 terms and to terms embedded in the sets on level 2 and level 3.

Presently, approximately 1,850 of the 13,500 controlled terms in GeoRef can be used on level 1 and/or level 2. The remainder are permitted on level 3 only. Uncontrolled terms are also permitted on level 3. These 1,850 terms appear in the Alphabetical Term List in this Guide.

For each kind of set, certain terms, or classes of terms, are prescribed. For instance, for an area set, the level-1 term must be one of the area terms in List O. The level-2 term is one of the discipline terms given in List B, such as "geochemistry" and "economic geology". And the level-3 term is a topic to be chosen from the topics appropriate for the level-2 term, e.g., for "economic geology", level 2, the topics on level 3 are commodities from List C. This close control of the sets results in uniformity in the indexing and the grouping of related documents in the index.

The principal sets and important changes which have occurred in the sets between 1969 and 1982 are given below. To locate specific topics described below, check the Index to this Guide. Capitalization in the examples below follows the usage of the year of the example. From 1969 to 1977, the first word on each level of the sets was capitalized; from 1978 on only proper nouns have been capitalized in the sets.

PRINCIPAL INDEX SETS
Areas

Since 1969, the first level of the area set which consists of a term from Special List O, Geographic Terms, has remained unchanged, except for addition of some new terms such as "Western Interior" and "West Germany". New terms are dated in the Alphabetical Term List. The second level has changed to limit it to geological disciplines. In the past, terms such as "absolute age" and "earthquakes" had been used on the second level as well. Starting in 1978, first-level area terms were automatically cross-referenced from commodities, ages and other important terms, e.g., earthquakes (see example below). These cross-references are included in the Alphabetical Term List.

1969:
   Delaware
      Sedimentary petrology
         Delaware bay, sediments
1970:
   Colorado
      Absolute age
         Tertiary, rhyolite, Denver basin

   Colorado
      Earthquakes
         Denver, fluid pressure
1973:
   Egypt
      Areal geology
         Stratigraphy, geochronology, geomorphology, Kom Ombo, Nile Valley
1980:
   earthquakes
      see also under seismology
      under Afghanistan, Alaska, etc.
1982:
   Aegean region
      tectonophysics plate tectonics: Reconstruction of stress fields for the Aegean in a finite element model

Igneous, Metamorphic and Sedimentary Rocks

These sets have remained unchanged on the first level, i.e., the first level is one of the following terms, "igneous rocks", "metamorphic rocks" or "sedimentary rocks". The second level began in 1969 having a specific rock name, but in 1972 the second level became more tightly controlled, and had either a topic or rock family. Some examples:

1969:
   Igneous rocks
      Tonalite
         Petrology, structure, Mexico

Metamorphic rocks
Pyroxene granulite
Absolute age, Sr/Rb

Sedimentary rocks
Arenite
Genesis, Ontario, Holleford meteorite
crater
1972:
Igneous rocks
Gabbro family
Anorthosite, gabbro, geochemistry,
Norway
Metamorphic rocks
Granulites
Eclogite, basalt, transition, mantle
Sedimentary rocks
Clastics, nonterrigenous
Diatomite, composition, carnotite,
Quaternary, Italy, Viterbo
1978:
igneous rocks
diorite family
diorite: The Argenta diorite, Utah; a
mineralized intrusive complex north of
the Uinta Axis
metamorphic rocks
mylonites
textures: Ductile deformation zones
and mylonites; the mechanical processes involved in the deformation of
crystalline basement rocks
sedimentary rocks
clastic rocks
petrography: Alteration of Proterozoic
sandstones from southern Ulutau during transition from epigenesis to early
stages of metamorphism
1980:
igneous rocks
ultramafic family
ophiolite: A petrologic and stable-iso-
tope study of the Oman ophiolite; implications for origin and metamorphism of the oceanic crust
metamorphic rocks
metaigneous rocks
metagabbro: Oceanic basaltic magmas in accretionary prisms; the Fran-
ciscan intrusive gabbros
sedimentary rocks
organic residues
resinite: Resinite; a potential petroleum source in the Upper Cretaceous
Terrtiary of the Beaufort-Mackenzie
Basin

Minerals

"Mineral data" was the first-level term for minerals from 1969 through 1971, when "minerals"
was added. "Mineral data" was still used in 1971
and 1972, but afterward no longer existed as a

first-level term. The second level-terms were spe-
cific minerals until 1972, when mineral groups
were used. Some examples:
1969:
Mineral data
Zircon
Absolute age, Guatemala
1971:
Mineral data
Topaz
Composition, USSR
Minerals
Spinel
Description, Australia
1972:
Minerals
Ring silicates
Axinite, crystal structure
1978:
minerals
sulfides
stannite: Geological features of tin-ore
deposits with high sulfostannate con-
tent

Commodities and Elements

The commodity and element sets changed very
little from 1969 to 1977; they could not be distin-
guished from one another by first-level terms, and
the second level was either an area term or
general term. In 1978, the second level became
more controlled and no area terms were included.
Commodities related to a specific area were
automatically cross-referenced to that area; elements were not. In 1981, parallel terms were
added to distinguish commodities from elements,
e.g., "copper ore" was added for copper as a
commodity and the term "copper" was restricted
to copper as an element.
1969:
Sulfur
Alberta
Panther river, possibilities
Sulfur
Geochemistry
Bottom sediments, Pacific Ocean
1978:
iron
see also under economic geology
under Africa, Angola, Austria, etc.
iron
exploration
ore deposits: Effectiveness of geophysical well-logging in the evaluation
of iron-ore deposits
1981:
copper
geochemistry
magmas: On the fractionation of sulfur, copper, and related transition elements in silicate liquids
copper ores

exploration
  magnetic methods: Some theoretical
  research and practical results of the
  magnetic induced polarization (MIP)
  method

Soils

From 1969 to 1971, this set remained relatively unchanged. The second-level term of this set was either a controlled general term or an area term. In 1972, an area could no longer be used on the second level. The term surveys was then introduced on the second level and the area term was then placed on the third level. Here are some examples:

1969:
  Soils
    Africa
      Rhodesian plateau
1972:
  Soils
    Surveys
      Asia, Formosa, clay mineralogy
1978:
  soils
    surveys
      Texas: Quality of irrigation return flow in the Mesilla Valley soils morphology profiles: Soil differences within spruce-fir forested and century-old burned areas of Libby Flats, Medicine Bow Range, Wyoming

Fossils

The fossil set remained the same from 1969 to 1971 with the first level being mostly phyla and classes of Mollusca. Second-level terms were taxonomic or ages. In 1972, the classes of Mollusca were moved down to second level. The second-level terms were either taxonomic or general terms. In 1981, parallel common-name terms were added for use in biostratigraphy papers and the systematic terms were restricted to paleontology papers, e.g., "foraminifers" was added for biostratigraphy and "foraminifera" was restricted to paleontology.

1969:
  Cephalopoda
    Jurassic
      Colombia, Kimmeridgian, Jipi formation, Guajira peninsula
1972:
  Mollusca
    Cephalopoda
      Carboniferous, Chesterian, Morrowan, Arkansas, Boston Mountains, new taxa, Ammonoidea
1978:
  Pisces
    Osteichthyes
      Cenozoic: Role of the Bering land bridge in the distribution of esocoid fishes
1982:
  foraminifers
    biostratigraphy
      Carboniferous: Lisburne Group, Cape Lewis—Niak Creek, northwestern Alaska

Ages

Until 1972, the age terms were the same as they had been in 1933 in the *Bibliography and Index of Geology Exclusive of North America*. In 1971, the Cenozoic epochs (e.g., Eocene and Pleistocene) became first-level terms. In 1978, the Tertiary periods (Paleogene and Neogene) and the Precambrian subdivisions (Archean and Proterozoic) became first-level terms. The second-level terms were either an area or general term until 1978, when the second-level term had to be a discipline or category term. If area-related, the age is now automatically cross-referenced to the area.

1969:
  Cretaceous
    Europe
      France-Belgium-England
1972:
  Oligocene
    Pacific Ocean
      Equatorial, Radiolaria, upper Oligocene
1978:
  Carboniferous
    see also under stratigraphy
      under Afghanistan, Alabama, etc.
1980:
  Quaternary
    sedimentary petrology
      calcrete: Some factors influencing Quaternary calcrete formation and distribution in alluvial fill of arid basins

Sediments and Sedimentary Structures

These sets remained unchanged from 1969 until 1974 when the second and third levels became much more controlled, i.e., the type of sediment or sedimentary structure was put on the second level followed by the specific sediment or sedimentary structure on the third. For example:

1969:
  Sediments
    Silt
      Distribution, Arctic Ocean, White Sea, dye tracers
  Sedimentary structures
    Cross-bedding
      Colorado, Devonian, west-central
1974:
  Sediments
    Clastics, terrigenous
      Silt, clay, laminated, lebensspuren, Quaternary, Europe, Austria

Sedimentary structures
Planar-bedding structures
Cross-bedding, clastics, Zambia
1978:
sediments
clastic sediments
ooze: Deep-sea ichnofacies; modern organism traces on and in pelagic carbonates of the western equatorial Pacific
sedimentary structures
bedding plane irregularities
current markings: Eddy markings from the Precambrian rocks of Kadana, Panchmahals, Gujarat, India

Disciplines

Discipline terms can be both first-level terms and second-level terms under areas. They have remained basically the same since 1969, but terms such as "tectonophysics" and "environmental geology" have been added. First-level usage has always been quite general for most discipline terms; however, some such as "engineering geology" and "environmental geology" have evolved from specific to general use, and terms such as "geomorphology" have remained specific.
1969:
Economic geology
Classification
Hydrocarbons, reserves
1974:
Engineering geology
Geologic hazards
Avalanches, USSR, Caucasus
Geomorphology
Fluvial features
Drainage patterns, Australia, South Australia
1978:
sedimentary petrology
history
1724–1974: The Academy of Sciences of the USSR and the development of lithology (on the Academy's 250th anniversary)
1981:
environmental geology
practice
review: Environmental geology
1982:
engineering geology
education
curricula: Measuring effectiveness of a college-level environmental earth-science course by changes in commitment to environmental issues

CROSS-REFERENCES

Since 1978 the Subject Index has included cross-referencing. This has been done by establishing a long list of cross-references which are added by computer, if appropriate. The method assures we miss no cross-references, eliminates blind cross-references, and has permitted us to add steadily to the list of cross-references.

The following types of cross-references are included:
–From a synonym or form not used in the system to an index term, e.g.:
genesis of ore deposits
see mineral deposits, genesis
Pelecypoda
see Bivalva *under* Mollusca
man, fossil
see fossil man
–From a general term, usually an area, to specific terms, e.g.:
Asia
see also Afghanistan; Arabian Peninsula; Bangladesh; . . .
–From a general term to a group of terms, e.g.:
fossils
see also under appropriate fossil group (which can be found in Special List F)
–From a term to the entries where the term appears on level 2 or level 3, e.g.:
Carboniferous
see also under stratigraphy
under Afghanistan; Alabama; Alaska; . . .
Coal
see also under economic geology
under Alabama; Alaska; Alberta; . . .
Magnetic field
see under Earth
andesite
see under andesite-rhyolite family
under igneous rocks
–From a term to related terms, e.g.:
lava
see also igneous rocks, magmas
mathematical geology
see also automatic data processing
Note: "*See*" is used for references from terms which are not valid level-1 terms. References from valid level-1 terms have "*see also*".

Introduction

This list consists of terms and cross-references found in the Subject Index of the *Bibliography and Index of Geology.* Its purpose is to guide you to the right terms to search in the Subject Index.

Terms—All terms used in level 1 or level 2 of the Subject Index are given, together with notes on how the terms are used. Most notes include a reference to the Special List where the term can be seen in the context of the broader, narrower and related terms of its group, and where there is an explanation of how index sets containing the term are structured. This provides you a glance at the rules followed by the indexers in using the term.

Only terms currently used are given, but references to previously used terms are sometimes included in the notes. Additional information on terms no longer in use can be found in Special List A.

Cross-references—Cross-references are added to the Subject Index by computer from a list of possible cross-references that we have compiled. All possible cross-references have been included herein. They can be spotted by the "CR" that precedes them. The cross-reference is given in a model form in which an underscore stands for "any term used in this position will be included in the cross-reference". For example, the cross-reference from "iron ores" reads:

see also under economic geology
  under———

This means that any time "iron ores" occurs in level 3 under "economic geology" on level 2 under any term on level 1 there will be a cross-reference such as the following from the May 1982 Bibliography:

iron ores *see also under* economic geology *under* Alaska; Argentina; Czechslovakia; Syria; U.S.S.R.; Western Australia

See Entries—These are from second or third words of compound index terms which appear in the Alphabetical Term List to the index terms. They are internal to this list and are not found as cross-references in the Bibliography. They are always introduced by the term "see". For example:

engineering *see* petroleum engineering

5. ALPHABETICAL TERM LIST

absolute age
For radiometric or radiogenic (isotopic) dating; for non-isotopic dating, see geochronology. Includes use on level 1; on level 2 under orogeny(1). If 1, term set options are:
    dates
    material [rock group, rock type, mineral name (e.g. charcoal, granite, metamorphic rocks, shells, sediments)]
    methods
    name of method [Ar/Ar, C-14, H-3, He-4/He-3, Io/Th, K/Ar, Pb-alpha, Pb/Pb, Pb/Th, Pb-210, Re/Os, Sr/Rb, Th/Th, Th/U, U/He, U/Pb, U-238/Pb-206, U-235/Pb-207, U/Th/Pb, uranium disequilibrium]
    techniques
    subtopic [e.g. instruments, models, sample preparation, sampling]
    topic [applications, bibliography, catalogs, interpretation, philosophy]
    subtopic
—— *CR see also* geochronology; isotopes

absorption
Includes use on level 2 under aurora(1); on level 3 under spectroscopy(1) methods(2); on level 3 under geochemistry(1) processes(2).

absorption and scattering
Includes use on level 2 under aeronomy(1).

abundance
Includes use on level 2 under chemical elements (list D) and under isotopes(1). Before 1981, also included use on level 2 under commodity terms. As of 1981, use only in the chemical sense.

Acantharina
Suborder. Includes use on level 2 under Radiolaria(1). See list F.

acidic composition
Term introduced in 1978. Includes use on level 2 under igneous rocks(1).

acids *see* amino acids; fatty acids; humic acids

Acoela
Term introduced in 1981. Includes use on level 2 under Mollusca(1). See list F.

acoustical logging
Not a valid index term from 1975 to 1977. After 1977, includes use on level 2 under well-logging(1).

acoustical methods
Includes use on level 2 under geophysical methods(1).

acoustical surveys *CR see under* geophysical surveys *under* ——

Acreodi
Term introduced in 1981. Includes use on level 2 under Mammalia(1). See list F.

acritarchs
Hystrichosphaeridae is included here and under Dinoflagellata. Includes use on level 2 under palynomorphs(1). See list F.

Actiniaria
Includes use on level 2 under Coelenterata(1). See list F.

Actinistia
Term introduced in 1981. Includes use on level 2 under Pisces(1). See list F.

actinium
Includes use on level 1 and 2 as a chemical element (list D).

Actinodontida
Term introduced in 1981. Includes use on level 2 under Mollusca(1). See list F.

activation analysis *CR see under* methods *under* chemical analysis

active layer
As of 1978, term is used on level 2 under permafrost(1).

Adriatic Sea
Between Italy on the W, and Albania and Yugoslavia on the E. Includes use on level 1 as an area term (list O). For term set options see list B.

Aegean Sea
Between Greece on the W and Turkey on the E. As of 1977 includes use on level 1 as an area term (list O). See list B for term set options.

aerial photography
As of 1978, term is used on level 2 under remote sensing(1).
—— *CR see under* remote sensing

aeromagnetic surveys *CR see* magnetic surveys *under* geophysical surveys *under* ——

aeronomy
Usually out-of-scope for GeoRef. Includes use on level 1. Term set options are:
    absorption and scattering
    subtopic
    composition
    element or compound
    densities and temperatures
    element or compound
    diffusion

    mechanism
    instruments
    name of instrument
    ionization
    latitude
    source of energy [e.g. cosmic rays, X-rays]
    techniques
    name of technique
    tides
    subtopic
    turbulence
    type
    waves
    type
    winds
    altitude by region [e.g. stratosphere, troposphere]

aerosols
Includes use on level 2 under meteorology(1).

Afars and Issas Territory *CR see* Djibouti

affinities
Includes use on level 2 or 3 under commodity terms (list C); used mostly on level 3 under fossil groups(1). See list F.

Afghanistan
Includes use on level 1 as an area term (list O). For term set options see list B.

Africa
Includes use on level 1 or 2 as an area term (list O). For term set options see list B.
—— *CR see also* Algeria; Angola; Benin; Botswana; Burundi; Cameroon; Cape Verde Islands; Central African Republic; Chad; Congo; Djibouti; Egypt; Equatorial Guinea; Ethiopia; Gabon; Gambia; Ghana; Guinea; Guinea-Bissau; Ivory Coast; Kenya; Lesotho; Liberia; Libya; Malagasy Republic; Malawi; Mali; Mauritania; Morocco; Mozambique; Namibia; Niger; Nigeria; Rwanda; Sahara; Senegal; Sierra Leone; Somali Republic; South Africa; Sudan; Swaziland; Tanzania; Togo; Tunisia; Uganda; Upper Volta; Zaire; Zambia; Zimbabwe

aftershocks
Includes use on level 2 under earthquakes(1).

age
Includes use as level 2 or 3 term appropriate to a large number of topics, e.g. on level 2 under ocean basins(1) and under ground water(1). See list G.

—— *see* absolute age; exposure age

ages *see* ancient ice ages

Agnatha
As of 1981, includes use on level 1 and 2 as a fossil term (list F). Before 1981, included use on level 2 under Pisces(1).

Agnostida
Includes use on level 2 under Trilobita(1). See list F.

agriculture
As of 1978, term is used on level 2 under land use(1).

air
As of 1978, term is used on level 2 under pollution(1).

Alabama
Includes use on level 1 as an area term (list O). For term set options see list B.

Alaska
Includes use on level 1 as an area term (list O). For term set options see list B.

Albania
Includes use on level 1 as an area term (list O). For term set options see list B.

albedo
Includes use on level 2 under ionosphere(1) for neutron albedo, magnetic albedo, and albedo of electromagnetic waves; on level 3 under interplanetary space(1) cosmic rays(2) and under magnetosphere(1) cosmic rays(2).

Alberta
Includes use on level 1 as an area term (list O). For term set options see list B.

algae
Includes use on level 1 and 2 as a fossil term (list F).

—— *see* calcareous algae

algal flora
Term introduced in 1981 as the common name for algae. See list F. To be used for algae when they are used as tools in stratigraphy. Includes use on level 1. If 1, term set options are:
  biostratigraphy
  age [list E]

Algeria
Includes use on level 1 as an area term (list O). For term set options see list B.

Alismidae
Term introduced in 1981. Includes use on level 2 under angiosperms(1). See list F.

alkali basalts
Term introduced in 1981. Includes use on level 2 under igneous rocks(1). See list H.

—— *CR see under* igneous rocks

alkali diorites
Term introduced in 1981. Includes use on level 2 under igneous rocks(1). See list H.

—— *CR see under* igneous rocks

alkali feldspar
As of 1981, includes use in combination with framework silicates (i.e. framework silicates, alkali feldspar) on level 2 under minerals(1). See list L. Before 1981, included use on level 3 under minerals(1) framework silicates, feldspar group(2).

alkali gabbros
Term introduced in 1981. Includes use on level 2 under igneous rocks(1). See list H.

—— *CR see under* igneous rocks

alkali granites
Term introduced in 1981. Includes use on level 2 under igneous rocks(1). See list H.

—— *CR see under* igneous rocks

alkali syenites
Term introduced in 1981. Includes use on level 2 under igneous rocks(1). See list H.

—— *CR see under* igneous rocks

alkalic amphibole
Term introduced in 1981. Includes use in combination with chain silicates (i.e. chain silicates, alkalic amphibole) on level 2 under minerals(1). See list L.

alkalic composition
Term introduced in 1978. Includes use on level 2 under igneous rocks(1).

alkalic pyroxene
Term introduced in 1981. Includes use in combination with chain silicates (i.e. chain silicates, alkalic pyroxene) on level 2 under minerals(1). See list L.

Allogromiina
Includes use on level 2 under foraminifera(1). See list F.

alloys
As of 1981, includes use on level 2 under minerals(1). See list L. y

—— *CR see under* minerals

alluvium *CR see under* clastic sediments *under* sediments

Alps
Great mountain system of S central Europe. Index countries as applicable. Includes use on level 1 as an area term (list O). For term set options see list B.

alteration
Formerly a general term appropriate to a large number of topics. As of 1976, use was restricted to any process having to do with changes in temperature and pressure, i.e. diagenesis, hydrothermal alteration,

metasomatism, weathering, etc. The term is used to denote the whole spectrum of chemical and physical change. Includes use on level 2 under igneous rocks(1).

aluminosilicates
As of 1981, includes use on level 2 under minerals(1). See list L. Before 1981, included use on level 3.

aluminum
Includes use on level 1 and 2 as a chemical element (list D). As of 1981, use aluminum ores for aluminum as a commodity. Before 1981, included use on level 1 as a commodity term.

aluminum ores
Term introduced in 1981. Includes use on level 1 and 2 as a commodity term (list C).

—— *CR see also under* economic geology *under* ——

Alveolinellidae
Includes use as level 2 term under foraminifera(1). See list F.

americium
Includes use on level 1 and 2 as a chemical element (list D).

amino acids
Includes use on level 2 under organic materials(1).

—— *CR see under* organic materials

Ammodiscacea
Includes use on level 2 under foraminifera(1). See list F.

Ammonoidea
As of 1981, includes use on level 2 under Mollusca(1). Before 1981, included use on level 3 under Mollusca(1) Cephalopoda(2). See list F.

Amphibia
Includes use on level 1 and 2 as a fossil term (list F).

amphibians
Term introduced in 1981 as the common name for Amphibia. See list F. To be used for Amphibia when they are used as tools in stratigraphy. Includes use on level 1. If 1, term set options are:
  biostratigraphy
  age [List E]

amphibole *see* alkalic amphibole

amphibole group
Includes use in combination with chain silicates (i.e. chain silicates, amphibole group) on level 2 under minerals(1). See list L.

amphibolites
Includes use on level 2 under metamorphic rocks(1). See list J.

analogs *see* systems analogs

analysis
Includes use on level 2 under name of element(1), under isotopes(1) and paleoecology(1). See list D. For

methods, see chemical analysis(1), spectroscopy(1), X-ray analysis(1), and thermal analysis(1). For data, see material name.

—— see activation analysis; chemical analysis; differential thermal analysis; environmental analysis; structural analysis; thermal analysis; thermogravimetric analysis; thermomagnetic analysis; vacuum fusion analysis

Anapsida
Includes use on level 2 under Reptilia(1). See list F.

Anarcestida
Term introduced in 1981. Includes use on level 2 under Mollusca(1). See list F.

ancient ice ages
Term introduced in 1978 on level 2 under glacial geology (1).

Andes
Mountain system extending along W coast from Venezuela and Colombia to Tierra del Fuego. Index countries as applicable. Includes use on level 1 as an area term (list O). For term set options see list B.

andesite *CR see under* —— under igneous rocks

andesites
Term introduced in 1981. Includes use on level 2 under igneous rocks(1). See list H.

—— *CR see under* igneous rocks

Andorra
A principality in E Pyrenees between France and Spain. Includes use on level 1 as an area term (list O). For term set options see list B.

angiosperm flora
Term introduced in 1982 as the common name for angiosperms. See list F. To be used for angiosperms when they are used as tools in stratigraphy. Includes use on level 1. If 1, term set options are:
biostratigraphy
age [list E]

angiosperms
Includes use on level 1 and 2 as a fossil term (list F).

Angola
Peoples Republic of Angola. Includes use on level 1 as an area term (list O). For term set options see list B.

Annelida
As of 1981, includes use on level 2 under worms(1). See list F. Before 1981, included use on level 1.

annual report
From 1978-1980, included use on level 2 under surveys(1). As of 1981, includes use on level 2 under survey organizations(1).

anomalies
Includes use as a level 2 or 3 term appropriate to a large number of topics, e.g. on level 2 under atmosphere(1), ecology(1), heat flow(1), and isostasy(1). See list G.

Antarctic Ocean
An arbitrary definition, considered the equivalent of Southern Ocean. Waters about Antarctica are sometimes called the Antarctic Ocean, but actually are only the parts of the Atlantic, Pacific and Indian Oceans south of approximately 55° S. Index oceans as applicable. Includes use on level 1 or 2 as an area term (list O). If 1, see term set options under list B.

Antarctica
Continent centering on the South Pole. Includes use on level 1 or 2 as an area term (list O). If 1, see term set options under list B.

Anthozoa
Includes use on level 2 under Coelenterata(1). See list F.

—— *CR see under* Coelenterata

antimonates
Term introduced in 1981. Includes use on level 2 under minerals(1). See list L.

—— *CR see under* minerals

antimonides
As of 1981, includes use on level 2 under minerals(1). See list L. Before 1981, included use on level 3 under minerals(1) sulfides(2).

—— *CR see under* sulfides *under* minerals

antimonites
Term introduced in 1981. Includes use on level 2 under minerals(1). See list L.

—— *CR see under* minerals

antimony
Includes use on level 1 and 2 as a chemical element (list D). As of 1981, use antimony ores for antimony as a commodity. Before 1981, included use on level 1 as a commodity term.

antimony ores
Term introduced in 1981. Includes use on level 1 and 2 as a commodity term (list C).

—— *CR see also under* economic geology *under* ——

Apennines
Mountain range extending the full length of the Italian peninsula. Includes use on level 1 as an area term (list O). For term set options see list B.

Aplacophora
Includes use on level 2 under Mollusca(1). See list F.

Appalachians
Mountain system of eastern North America extending from Canadian maritime provinces to Alabama. Index countries as applicable. Includes use on level 1 as an area term (list O). For term set options see list B.

applications
Includes use as level 2 or 3 term appropriate to a large number of topics, e.g. on level 2 under mineral exploration(1). See list G.

applied geophysics *CR see* geophysical surveys

aquifers
Includes use on level 2 under ground water(1); on level 3 under hydrogeology(1).

—— *CR see under* ground water

Arabian Peninsula
Between Red Sea on the W; and the Persian Gulf, Gulf of Oman, and Arabian Sea on the E. Index countries as applicable. Includes use on level 1 and 2 as an area term (list O). For term set options see list B.

—— *CR see also* Bahrain; Kuwait; Oman; Qatar; Saudi Arabia; Southern Yemen; United Arab Emirates; Yemen

Arabian Sea
Between the Arabian Peninsula on the W and Pakistan and India in the E. As of 1977, includes use as a level 1 area term (list O). For term set options see list B.

—— *CR see also* Gulf of Aden

Arachnida
Class. Includes use on level 2 under Arthropoda(1). See list F.

Araeoscelidia
Term introduced in 1981. Includes use on level 2 under Reptilia(1). See list F.

Archaeocyatha
Includes use on level 1 and 2 as a fossil term (list F).

Archaeogastropoda
As of 1981, includes use on level 2 under Mollusca(1). See list F. Before 1981, included use on level 3 under Mollusca(1) Gastropoda(2).

archaeology *CR see also under* stratigraphy *under* ——

Archaeornithes
Includes use on level 2 under Aves(1). See list F.

Archean
Includes use on level 1 and 2 as an age term (list E) as of 1978. As of 1981, used for lower Precambrian.

—— *CR see also* Precambrian; *see also under* geochronology *under* ——; *see also under* stratigraphy *under* ——

Archeocopida
Order. Includes use on level 2 under Ostracoda(1).

archeology *CR see* archaeology *under ____ under ____*

Archosauria
Includes use on level 2 under Reptilia(1). See list F. y

Arcina
Term introduced in 1981. Includes use on level 2 under Mollusca(1). See list F.

Arctic Islands *CR see* Arctic Archipelago

Arctic Ocean
Extends from North Pole southward to approximately 70° N and is bounded by Alaska, Canada, Greenland, Norway, and the USSR. Includes use on level 1 or 2 as an area term (list O). If 1, see term set options under list B.

Arctic region
The Arctic Ocean and islands in it and adjacent to it plus mainland surfaces N of the Arctic circle. Index Alaska, Arctic Archipelago, Arctic Ocean, Greenland, and countries as applicable. Includes use on level 1 or 2 as an area term (list O). If 1, see term set options under list B.
—— *CR see also* Greenland; Jan Mayen; Spitsbergen

Arctocyonia
Term introduced in 1981. Includes use on level 2 under Mammalia(1). See list F.

areal geology
Used for entries that might properly be placed under three or more of the second-order headings such as geomorphology, stratigraphy, structural geology. Includes use on level 2 under area terms(1). See list B.

areal studies
Includes use on level 2 under clay mineralogy(1).

Arecidae
Term introduced in 1981. Includes use on level 2 under angiosperms(1). See list F.

Argentina
Includes use on level 1 as an area term (list O). For term set options see list B.

argon
Includes use on level 1 and 2 as a chemical element (list D).

arid environment
Term introduced in 1978. Includes use on level 2 under land use(1).

Arizona
Includes use on level 1 as an area term (list O). For term set options see list B.

Arkansas
Includes use on level 1 as an area term (list O). For term set options see list B.

arsenates
Includes use on level 2 under minerals(1). See list L.
—— *CR see under* minerals

arsenic
Includes use on level 1 and 2 as a chemical element (list D). As of 1981, use arsenic ores for arsenic as a commodity. Before 1981, included use on level 1 as a commodity term.

arsenic ores
Term introduced in 1981. Includes use on level 1 and 2 as a commodity term (list C).
—— *CR see also under* economic geology *under ____*

arsenides
As of 1981, includes use on level 2 under minerals(1). See list L. Before 1981, included use on level 3 under minerals(1) sulfides(2).
—— *CR see under* sulfides *under* minerals

arsenites
Includes use on level 2 under minerals(1). See list L. y
—— *CR see under* minerals

artesian waters
Includes use on level 2 under ground water(1).
—— *CR see under* ground water

Arthropoda
Includes use on level 1 and 2 as a fossil term (list F).

Articulata
Includes use on level 2 under Brachiopoda(1). See list F.

Artiodactyla
Includes use on level 2 under Mammalia(1). See list F.

asbestos deposits
Term introduced in 1981. Includes use on level 1 and 2 as a commodity term (list C).
—— *CR see also under* economic geology *under ____*

Asia
Includes use on level 1 or 2 as an area term (list O). If 1, see term set options under list B.
—— *CR see also* Afghanistan; Arabian Peninsula; Bangladesh; Bhutan; Burma; Cambodia; China; Far East; Himalayas; Hong Kong; India; Indochina; Indonesia; Iran; Japan; Korea; Laos; Malay Archipelago; Malaysia; Mongolia; Nepal; Pakistan; Philippine Islands; Singapore; Sri Lanka; Taiwan; Thailand; Vietnam

asphalt *CR see under* bitumens *under* organic materials

assemblages *see* mineral assemblages

associations
Includes use on level 1 (list A). Reserved for organizations not included under museums or survey organizations. Term set options are:
  topic [List B (except areal geology), or extraterrestrial geology, general, geophysics]
  name of group, conference, meeting, symposium [conventional abbreviations for well known organizations may be used, e.g. AAPG, AGI, CNRS, COMECON, CSIRO, EURATOM, GSA, INQUA, IUGS, NASA, NATO, NOAA, UNESCO, VINITI]

Astartida
Term introduced in 1981. Includes use on level 2 under Mollusca(1). See list F.

Asteridae
Term introduced in 1981. Includes use on level 2 under angiosperms(1). See list F.

Asteroidea
As of 1981, includes use on level 2 under Echinodermata(1). See list F. Before 1981, included use on level 3 under Echinodermata(1) Stelleroidea(2).

asteroids
Includes use on level 1 (list A). Term set options are:
  topic [composition, cosmochemistry, distribution, evolution, genesis, observations, properties, theoretical studies]
  (name of asteroid) or subtopic

Astrapotheria
Includes use on level 2 under Mammalia(1). See list F.

astrophysics *CR see under ____ under* planetology

astrophysics and solar physics
Usually out-of-scope for GeoRef. Includes use on level 1. Term set options are:
  corona
    subtopic
  electromagnetic radiation
    type of radiation or subtopic
  solar flares
    type of radiation or subtopic
  instruments
    name of instrument
    phenomena measured
  magnetic field
    subtopic
  particle radiation
    name of particle or subtopic
  surface phenomena
    name of phenomena or subtopic

Atlantic Coastal Plain
Index states as applicable. Includes use on level 1 as an area term (list O). For term set options see List B.

Atlantic Ocean
Includes use on level 1 or 2 as an area term (list O). If 1, see term set options under list B.
—— CR see also Baltic Sea; Caribbean Sea; Celtic Sea; English Channel; Gulf of Mexico; Irish Sea; North Sea

Atlantic Ocean Islands
Term introduced in 1981. Includes use on level 1 as an area term (list O). For term set options, see list B.
—— CR see also Azores; Bermuda; Canary Islands; Cape Verde Islands; Faeroe Islands; Falkland Islands; Iceland; Jan Mayen; Madeira; Shetland Islands

Atlantic region
Term introduced in 1976. An artificial term used to indicate the coastal region immediately adjacent to the Atlantic Ocean. Includes land and the immediate littoral zone. Includes use on level 1 and 2 as an area term (list O). For term set options see list B.

atmosphere
Includes use on level 1(only for Earth); on level 2 or 3 under Moon(1) or other planets; on level 3 under environmental geology(1) pollution(2). For studies dealing with geochemical changes (short-term or during geologic time), observations related to interactions with phenomena of the Earth's surface, and other geophysical aspects. For changes in climate, see paleoclimatology. If 1, term set options are:
topic [age, changes, circulation, composition, evolution, general, genesis, instruments, observations, processes, properties, temperature]
subtopic

atmospheric precipitation
As of 1978, term is used on level 2 under hydrology(1).

aurora
Usually out-of-scope for GeoRef. Includes use on level 1 for special bibliographies; on level 2 under ionosphere(1). If 1, term set options are:
absorption
subtopic
auroral zone (including auroral oval)
subtopic
electrical field
topic [arcs, currents, electrojet, particle precipitation]
magnetic field
topic [currents, electrojet, magnetic storms, micropulsations, type of aurora]

morphology
motions or subtopic
orientation or subtopic
synoptic studies or subtopic
time variations or subtopic
observations
type of aurora
theoretical studies
topic

auroral zone
Including auroral oval. Includes use on level 2 under aurora(1).

Australasia
Islands of the South Pacific including Australia, New Zealand, and Papua New Guinea. Index countries as applicable. Includes use on level 1 or 2 as an area term (list O). If 1, see term set options under list B.
—— CR see also New Zealand; Papua New Guinea

Australia
Includes use on level 1 or 2 as an area term (list O). If 1, see term set options under list B.
—— CR see also New South Wales; Northern Territory; Queensland; South Australia; Tasmania; Victoria; Western Australia

Austria
Includes use on level 1 as an area term (list O). For term set options see list B.

automatic data processing
Includes use on level 1 and 2 (list A) for computer applications and machine modeling of geological data. Term set options are:
methods
name of method or application
topic [List B (except areal geology), or extraterrestrial geology, general, geophysics]
subtopic related to 2nd level term

autometamorphism
Includes use on level 2 under metamorphism(1).

avalanches
As of 1978, term is used on level 2 under geologic hazards(1).
—— CR see under —— under geomorphology; see under geologic hazards

Aves
Includes use on level 1 and 2 as a fossil term (list F).

Azores
Island group belonging to Portugal in E central North Atlantic Ocean. Includes use on level 1 as an area term (list O). For term set options see list B.

bacteria
Schizomycetes included here. Includes use on level 1 and 2 as a fossil term (list F); on level 3 under soils(1) biota(2) and under geochemistry(1).

Bactritida
Term introduced in 1981. Includes use on level 2 under Mollusca(1). See list F.

Bahamas
A chain of islands SE of Florida and N of Cuba. Includes use on level 1 as an area term (list O). For term set options see list B.

Bahrain
An archipelago in the W Persian Gulf. Includes use on level 1 as an area term (list O). For term set options see list B.

Bairdiomorpha
Term introduced in 1981. Includes use on level 2 under Ostracoda(1). See list F.

Balearic Islands
Off E coast of Spain. Forms Spanish province of Baleares. Includes use on level 1 as an area term (list O). For term set options see list B.

Balkan Peninsula
SE Europe between the Adriatic and Ionian Seas on the W; the Mediterranean Sea on the S; and the Aegean and Black seas on the E. Includes use on level 1 as an area term (list O). For term set options see list B.

Baltic region
Index Baltic Sea, countries and Soviet republics as applicable. Includes use on level 1 or 2 as an area term (list O). If 1, see term set options under list B.

Baltic Sea
Enclosed by Denmark, Sweden, Finland, Soviet Union, Poland, and Germany. Includes use on level 1 as an area term (list O). For term set options see list B.

Bangladesh
Formerly East Pakistan. Includes use on level 1 as an area term (list O). For term set options see list B.

Barbados
Easternmost island of the Lesser Antilles. Includes use on level 1 as an area term (list O). For term set options see list B.

barite deposits
Term introduced in 1981. Includes use on level 1 and 2 as a commodity term (list C).
—— CR see also under economic geology under ——

barium
Includes use on level 1 and 2 as a chemical element (list D).

barium feldspar
Term introduced in 1981. Includes use in combination with framework silicates (i.e. framework silicates, barium feldspar) on level 2 under minerals(1). See list L. Before 1981,

included use on level 3 under minerals(1) framework silicates, feldspar group(2).

barometry *see* geologic barometry

barrier islands
As of 1978, term is used on level 2 under shorelines(1).

Barytherioidea
Term introduced in 1981. Includes use on level 2 under Mammalia(1). See list F.

basalt *CR see under* ____ *under* igneous rocks

basalts
Term introduced in 1981. Includes use on level 2 under igneous rocks(1). See list H.
—— *CR see under* igneous rocks
—— *see* alkali basalts

base metals
As of 1978, term includes use on level 1 and 2 as a commodity (list C).
—— *CR see also under* economic geology *under* ____

Basin and Range Province
Level 1 area term as of 1978. Physiographic province in W and SW United States characterized by a series of tilted fault blocks forming longitudinal, asymmetric ridges of mountains and broad intervening basins. Index states as applicable.

basins *see* ocean basins

Basomatophora
Term introduced in 1981. Includes use on level 2 under Mollusca(1). See list F.

batholiths
Includes use on level 2 under intrusions(1).
—— *CR see under* intrusions

bauxite
Includes use on level 1 and 2 as a commodity term (list C). As of 1981, also includes use on level 3 under sedimentary rocks(1) clastic rocks(2). See list I.
—— *CR see also under* economic geology *under* ____

beaches
As of 1978, term is used on level 2 under shorelines(1).
—— *CR see under* shorelines; *see under* environment *under* conservation; *see under* shore features *under* geomorphology

bedding *see* graded bedding; planar bedding structures

bedding plane irregularities
Includes use on level 2 under sedimentary structures(1). See list K.

Belemnoidea
As of 1981, includes use on level 2 under Mollusca(1). See list F. Before 1981, included use on level 3 under Mollusca(1) Cephalopoda(2).

Belgium
Includes use on level 1 as an area term (list O). For term set options see list B.

Belize
Formerly British Honduras. Includes use on level 1 as an area term (list O). For term set options see list B.

Bellerophontina
Term introduced in 1981. Includes use on level 2 under Mollusca(1). See list F.

Benin
Was established in 1960 as Dahomey. Includes use on level 1 as an area term (List O). For term set options see list B.

Bennettitales
Includes use on level 2 under gymnosperms(1). See list F.

bentonite deposits
Term introduced in 1981. Includes use on level 1 and 2 as a commodity term (list C).
—— *CR see also under* economic geology *under* ____

Bering Sea
Between NE Siberia and Alaska with the Aleutian Islands on the S and Bering Strait on the N. Includes use on level 1 as an area term (list O). For term set options see list B.

Bermuda
British colony comprising over 300 coral islands about 640 miles ESE of Cape Hatteras. Includes use on level 1 as an area term (list O). For term set options see list B.

beryllium
Includes use on level 1 and 2 as a chemical element (list D). As of 1981, use beryllium ores for beryllium as a commodity. Before 1981, included use on level 1 as a commodity term.

beryllium ores
Term introduced in 1981. Includes use on level 1 and 2 as a commodity term (list C).
—— *CR see also under* economic geology *under* ____

Beyrichicopina
Term introduced in 1981. Includes use on level 2 under Ostracoda(1). See list F.

Bhutan
Kingdom in E Himalayas between Tibet and NE India. Includes use on level 1 as an area term (list O). For term set options see list B.

bibliography
Includes use on level 1 (list A); on level 2 under major disciplines; as a level 2 or 3 term appropriate to a large number of topics (list G). Covers major bibliographies on topics, areas or persons. If 1, term set options are:

topic [List B or extraterrestrial geology, general, geophysics] subtopic
—— *CR see also under* areal geology *under* ____; *see also under* economic geology *under* ____; *see also under* engineering geology *under* ____; *see also under* environmental geology *under* ____; *see also under* general *under* ____; *see also under* geochemistry *under* ____; *see also under* geochronology *under* ____; *see also under* geomorphology *under* ____; *see also under* geophysical surveys *under* ____; *see also under* hydrogeology *under* ____; *see also under* mineralogy *under* ____; *see also under* oceanography *under* ____; *see also under* paleobotany *under* ____; *see also under* paleontology *under* ____; *see also under* petrology *under* ____; *see also under* sedimentary petrology *under* ____; *see also under* seismology *under* ____; *see also under* soils *under* ____; *see also under* stratigraphy *under* ____; *see also under* structural geology *under* ____; *see also under* tectonophysics *under* ____

biocenoses
Includes use on level 2 and 3 under fossil group(1). See list F.

biochemistry
Includes use on level 2 under fossil group(1). See list F.

biocirculation
Includes use on level 2 under ocean circulation(1).

biogenic structures
Includes use on level 2 under sedimentary structures(1). See list K.

biogeochemical methods
Includes use on level 2 under mineral exploration(1).

biogeography
Includes use on level 1 (list A); on level 2 under fossil groups (list F). To be used for descriptions of the geographic distribution of both fossil and modern organisms. If 1, term set options are:
age (List E; only where one age is discussed)
  area (or global)
fossil group [List F]
  age
topic [catalogs, concepts, distribution, evolution, interpretation, nomenclature, observations, patterns]
  subtopic
—— *CR see also under* paleobotany *under* ____; *see also under* paleontology *under* ____; *see also under* stratigraphy *under* ____

biography
Includes use on level 1 (list A). This set is used only where there is an extensive bibliography. Term set options are:
general
  name of individual (last, first),
—— *CR see also* bibliography

biostratigraphy
As of 1981, includes use on level 2 under common names of fossil groups(1). See list F. Before 1981, included use on level 2 under scientific names of fossil groups(1).

biota
Includes use on level 2 under soils(1).

birds
Term introduced in 1981 as the common name for Aves. See list F. To be used for Aves when they are used as tools in stratigraphy. Includes use on level 1. If 1, term set options are:
biostratigraphy
age [List E]

bismuth
Includes use on level 1 and 2 as a chemical element (list D). As of 1981, use bismuth ores for bismuth as a commodity. Before 1981, included use on level 1 as a commodity term.

bismuth ores
Term introduced in 1981. Includes use on level 1 and 2 as a commodity term (list C).
—— *CR see also under* economic geology *under* ——

bismuthides
As of 1981, includes use on level 2 under minerals(1). See list L. Before 1981, included use on level 3 under minerals(1) sulfides(2).
—— *CR see under* sulfides *under* minerals

bitumens
Includes use on level 1 as a commodity term (list C); on level 2 under organic materials(1).
—— *CR see also under* economic geology *under* ——

Bivalvia
Includes use on level 2 under Mollusca(1). See list F.

Black Sea
Large inland sea bounded on the N and NE by the USSR, S by Turkey, and W by Bulgaria and Romania. Includes use on level 1 as an area term (list O). For term set options see list B.

Blastoidea
Includes use on level 2 under Echinodermata(1). See list F.

Blattopteroida
Term introduced in 1981. Includes use on level 2 under Insecta(1). See list F.

Bolivia
Includes use on level 1 as an area term (list O). For term set options see list B.

bonding
As of 1978, term is used on level 2 under crystal chemistry(1) and crystal structure(1).

book reviews
Includes use on level 1. Term set options are:
topic [List B, extraterrestrial geology, general, geophysics]
title

borates
Includes use on level 2 under minerals(1). See list L. As of 1981, use borate deposits for borates as a commodity. Before 1981, included use on level 3 as a commodity term.
—— *CR see under* minerals

boron
Includes use on level 1 and 2 as a chemical element (list D). As of 1981, use boron deposits for boron as a commodity. Before 1981, included use on level 1 as a commodity term.

boron deposits
Term introduced in 1981. Includes use on level 1 and 2 as a commodity term (list C).
—— *CR see also under* economic geology *under* ——

Botswana
Before 1966 was British protectorate of Bechuanaland. Includes use on level 1 as an area term (list O). For term set options see list B.

bottom features
Includes use on level 2 under ocean floors(1) and marine geology(1); on level 3 under oceanography(1).

boudinage *CR see under* —— *under* lineation

boundary interactions
Includes use on level 2 under meteorology(1).

boundary layer
Includes use on level 2 under ocean circulation(1).

bow shock waves
Includes use on level 2 under magnetosphere(1).

Brachiopoda
Includes use on level 1 and 2 as a fossil term (list F).

brachiopods
Term introduced in 1981 as the common name for Brachiopoda. See list F. To be used for Brachiopoda when they are used as tools in stratigraphy. Includes use on level 1. If 1, term set options are:

biostratigraphy
age [List E]

Brachiopterygii
Term introduced in 1981. Includes use on level 2 under Pisces(1). See list F.

Branchiopoda
Includes use on level 2 under Arthropoda(1). See list F. As of 1981, used for Conchostraca.

Brazil
Includes use on level 1 as an area term (list O). For term set options see list B.

breaking waves
Term introduced in 1978. Includes use on level 2 under ocean waves(1).

bridges
As of 1978, term is used on level 2 under foundations(1).
—— *CR see under* foundations

brines
Includes use on level 1 and 2 as a commodity term (list C).
—— *CR see also* bromine; *see also under* economic geology *under*
——

British Columbia
Includes use on level 1 as an area term (list O). For term set options see list B.

bromides
Term introduced in 1981. Includes use on level 2 under minerals(1). See list L.

bromine
Includes use on level 1 and 2 as a chemical element (list D). As of 1981, use bromine deposits for bromine as a commodity. Before 1981, included use on level 1 as a commodity term.

bromine deposits
Term introduced in 1981. Includes use on level 1 and 2 as a commodity term (list C).
—— *CR see also under* economic geology *under* ——

brown coal *CR see* lignite

bryophytes
Mosses. Includes use on level 1 and 2 as a fossil term (list F).

Bryozoa
Includes use on level 1 and 2 as a fossil term (list F).

bryozoans
Term introduced in 1981 as the common name for Bryozoa. See list F. To be used for Bryozoa when they are used as tools in stratigraphy. Includes use on level 1. If 1, term set options are:
biostratigraphy
age [List E]

buildings
As of 1978, term is used on level 2 under foundations(1).

—— *CR see under* foundations

Bulgaria
Includes use on level 1 as an area term (list O). For term set options see list B.

Buliminacea
Includes use on level 2 under foraminifera(1). See list F.

burial metamorphism
Term introduced in 1978. Includes use on level 2 under metamorphism(1).

Burma
Includes use on level 1 as an area term (list O). For term set options see list B.

bursts *see* rock bursts

Burundi
Formerly Urundi which was part of Belgian trust territory of Uranda-Urundi. Includes use on level 1 as an area term (list O). For term set options see list B.

cadmium
Includes use on level 1 and 2 as a chemical element (list D). As of 1981, use cadmium ores for cadmium as a commodity. Before 1981, included use on level 1 as a commodity term.

cadmium ores
Term introduced in 1982. Includes use on level 1 and 2 as a commodity term (list C).

calc-alkalic composition
Term introduced in 1978. Includes use on level 2 under igneous rocks(1).

calcareous algae *CR see under* —— *under* algae

calcic composition
Term introduced in 1978. Includes use on level 2 under igneous rocks(1).

Calcispongea
Includes use on level 2 under Porifera(1). See list F.

calcite *CR see also under* —— *under* minerals

calcite deposits
Term introduced in 1981. Includes use on level 1 and 2 as a commodity term (list C) for optical calcite.

—— *CR see also under* economic geology *under* ——

calcium
Includes use on level 1 and 2 as a chemical element (list D).

California
Includes use on level 1 as an area term (list O). For term set options see list B.

californium
Includes use on level 1 and 2 as chemical element (list D).

caliper logging
As of 1978, term is used on level 2 under well-logging(1).

Camaroidea
Includes use on level 2 under Graptolithina(1). See list F.

Cambodia
Khmer Republic or Kampuchea. Includes use on level 1 as an area term (list O). For term set options see list B.

Cambrian
Includes use on level 1 as an age term (list E); on level 2 under paleoterms, e.g. paleoecology, paleomagnetism, paleogeography. Above Precambrian, below Ordovician.

—— *CR see also under* geochronology *under* ——; *see also under* stratigraphy *under* ——

Cameroon
Includes former French trust territory and British trust territory of Southern Cameroons. Includes use on level 1 as an area term (list O). For term set options see list B.

Camptostromatoidea
Includes use on level 2 under Echinodermata(1). See list F.

Canada
Includes use on level 1 or 2 as an area term (list O). If 1, see list B for term set options.

—— *CR see also* Alberta; Appalachians; Arctic Archipelago; Atlantic Coastal Plain; British Columbia; Canadian Shield; Great Lakes region; Great Plains; Labrador; Manitoba; Maritime Provinces; New Brunswick; Newfoundland; Northwest Territories; Nova Scotia; Ontario; Prince Edward Island; Quebec; Rocky Mountains; Saskatchewan; Yukon Territory

Canadian Shield
Index states and provinces as applicable. Introduced as a level 1 area term in 1978.

canals
As of 1978, term is used on level 2 under waterways(1).

Canary Islands
Spanish controlled island group off NW Africa. Includes use on level 1 as an area term (list O). For term set options see list B.

canyons *see* submarine canyons

Cape Verde Islands
Former Portuguese overseas province W of Dakar, Senegal. Includes use on level 1 as an area term (list O). For term set options see list B.

Captorhinomorpha
Term introduced in 1981. Includes use on level 2 under Reptilia(1). See list F.

carbides
As of 1981, includes use on level 2 under minerals(1). See list L. Before 1981, included use on level 3 under minerals(1) native elements and alloys(2).

—— *CR see under* native elements and alloys *under* minerals

carbohydrates
Includes use on level 2 under organic materials(1).

—— *CR see under* organic materials

carbon
Includes use on level 1 and 2 as a chemical element (list D).

carbonate *see* sodium carbonate

carbonate rocks
Includes use on level 2 under sedimentary rocks(1). See list I.

—— *CR see under* sedimentary rocks

carbonate sediments
Includes use on level 2 under sediments(1). See list N.

carbonates
Includes use on level 2 under minerals(1). See list L.

—— *CR see under* minerals

carbonatites
Term introduced in 1981. Includes use on level 2 under igneous rocks(1). See list H.

—— *CR see under* igneous rocks

Carboniferous
Includes use on level 1 as an age term (list E); on level 2 under paleoterms, e.g. paleoecology, paleogeography, paleomagnetism. Above Devonian, below Permian.

—— *CR see also* Mississippian; Pennsylvanian; *see also under* geochronology *under* ——; *see also under* stratigraphy *under* ——

Carditida
Term introduced in 1981. Includes use on level 2 under Mollusca(1). See list F.

Caribbean region
Index Antilles, Caribbean Sea, Central America, Colombia, and Venezuela as applicable. Includes use on level 1 or 2 as an area term (list O). If 1, see list B for term set options.

Caribbean Sea
Bounded by the West Indies on the N and E, South America on the S, and Central America on the W. Includes use on level 1 or 2 as an area term (list O). If 1, see list B for term set options.

Carnivora
Includes use on level 2 under Mammalia(1). See list F.

Carpathians
Mountain system of E and central Europe enclosing the Alfold. Index Ukraine and countries as applicable. Includes use on level 1 as an area term (list O). For term set options see list B.

Carterinacea
Includes use on level 2 under foraminifera(1). See list F.

cartography
For methods, instruments, programs. Includes use on level 2 under maps(1).

Caryophyllidae
Term introduced in 1981. Includes use on level 2 under angiosperms(1). See list F.

case studies
As of 1978, term is used on level 2 under pollution(1), rock mechanics(1), and soil mechanics(1).

Caspian Sea
Largest inland sea in the world. Index Iran and/or the USSR as applicable. Includes use on level 1 as an area term (list O). For term set options see list B.

Cassidulinacea
Includes use on level 2 under foraminifera(1). See list F.

cataclasites
Includes use on level 2 under metamorphic rocks(1). See list J.

catalogs
Includes use on level 1 (list A); as a level 2 or 3 term appropriate to a large number of topics, e.g. on level 2 under paleontology(1). See list G. If 1, level 2 terms are:
topic [list B, extraterrestrial geology, general, geophysics]
subtopic (e.g. language if not English)

catastrophes
Term introduced in 1978. Includes use on level 2 under geologic hazards(1).

catastrophic waves
Term introduced in 1978. Includes use on level 2 under ocean waves(1).

causes
Includes use as level 2 or 3 term appropriate to a large number of topics, e.g. on level 2 under orogeny(1). See list G.

caves *CR see under* environment *under* conservation; *see under* solution features *under* geomorphology

Cayman Islands
Term introduced in 1981. Includes use on level 1 as an area term (list O).

Caytoniales
Includes use on level 2 under gymnosperms(1). See list F.

Celebes Sea
Bounded on N by Philippines, W by Borneo, and S by island of Celebes. Includes use on level 1 as an area term (list O). For term set options see list B.

cell dimensions
As of 1978, term is used on level 2 under crystal structure(1).

Cenozoic
Includes use on level 1 as an age term (list E); on level 2 under paleo-terms, e.g. paleoecology, paleogeography, paleomagnetism.
—— *CR see also* Eocene; Holocene; Miocene; Neogene; Oligocene; Paleocene; Paleogene; Pleistocene; Pliocene; Quaternary; Tertiary; *see also under* geochronology *under* ——; *see also under* stratigraphy *under* ——

Central African Republic
Formerly French territory of Ubangi-Shari. Became independent in 1960. Includes use on level 1 as an area term (list O). For term set options see list B.

Central America
Index countries as applicable. Includes use on level 1 or 2 as an area term (list O). If 1, see list B for term set options.
—— *CR see also* Belize; Costa Rica; El Salvador; Guatemala; Honduras; Nicaragua; Panama

Cephalodiscida
Includes use on level 2 under Pterobranchia(1). See list F.

Cephalopoda
Includes use on level 2 under Mollusca(1). See list F.

ceramic materials
Includes use on level 1 as a commodity term (list C).
—— *CR see also* clays; kaolin; *see also under* economic geology *under* ——

Ceratitida
Term introduced in 1981. Includes use on level 2 under Mollusca(1). See list F.

Ceratomorpha
Term introduced in 1981. Includes use on level 2 under Mammalia(1). See list F.

Ceriantipatharia
Includes use on level 2 under Coelenterata(1). See list F.

cerium
Includes use on level 1 and 2 as a chemical element (list D). As of 1981, use cerium ores for cerium as a commodity. Before 1981, included use on level 1 as a commodity term.

cerium ores
Term introduced in 1982. Includes use on level 1 and 2 as a commodity term (list C).

cesium
Includes use on level 1 and 2 as a chemical element (list D).

Cetacea
Includes use on level 2 under Mammalia(1). See list F.

Ceylon *CR see* Sri Lanka

Chad
Includes use on level 1 as an area term (list O). For term set options see list B.

Chaetognatha
Includes use on level 2 under worms(1). See list F.

chain silicates
Includes use on level 2 under minerals(1); in combination with amphibole group, alkalic amphibole, clinoamphibole, orthoamphibole, pyroxene group, alkalic pyroxene, clinopyroxene, and orthopyroxene (i.e. chain silicates, amphibole group) to form terms on level 2 under minerals(1). See list L.
—— *CR see under* minerals

changes
Includes use as level 2 or 3 term appropriate to a large number of topics, e.g. on level 2 under paleo-ecology(1). See list G.

changes of level
Used mostly under stratigraphy for sea-level (sometimes lake-level) changes of the upper Cenozoic. Includes use on level 1(list A). Term set options are:
topic [age, causes, concepts, correlation, detection, evolution, genesis, interpretation, mechanism, observations, patterns, processes, rates]
subtopic (no area term)
—— *CR see also* isostasy; *see also under* geomorphology *under* ——; *see also under* stratigraphy *under* ——

channels
As of 1978, term is used on level 2 under waterways(1).

Charophyta
Includes use on level 2 under algae(1). See list F.

Cheilostomata
Includes use on level 2 under Bryozoa(1). See list F.

Chelicerata
Subphylum. Includes use on level 2 under Arthropoda(1). See list F.

Chelonia
As of 1981, includes use on level 2 under Reptilia(1). See list F. Before 1981, included use on level 3 under Reptilia(1) Anapsida(2).

chemical analysis
Use only for methodology. Includes use on level 2 under minerals(1). For data, see appropriate material. Includes use on level 1 (list A). Term set options are:
methods
name of method [activation analysis, chromatography, colorimetry, electrolytic analysis, infrared spectroscopy, major-element analyses, minor-element analyses, polarography, spectroscopy, trace-element analyses, vacuum fusion analysis, volumetric analysis, wet methods, X-ray fluorescence]
techniques
subtopic [e.g. sample preparation, titration, material, reagent, etc.]

chemical explosions
As of 1978, term is used on level 2 under explosions(1).

chemically precipitated rocks
Includes use on level 2 under sedimentary rocks(1). See list I.

chemistry *see* crystal chemistry

Chile
Includes use on level 1 as an area term (list O). For term set options see list B.

China
Peoples Republic of China. Includes use on level 1 as an area term (list O). For term set options see list B.

Chiroptera
Includes use on level 2 under Mammalia(1). See list F.

Chitinozoa
Includes use on level 2 under palynomorphs(1). See list F.

—— *CR see under* palynomorphs

chlorides
As of 1981, includes use on level 2 under minerals(1). See list L. Before 1981, included use on level 3 under minerals(1) halides(2).

—— *CR see under* halides *under* minerals

chlorine
Includes use on level 1 and 2 as a chemical element (list D).

chlorite group
Includes use in combination with sheet silicates (i.e. sheet silicates, chlorite group) on level 2 under minerals(1). See list L.

Chlorophyceae
Term introduced in 1981. Includes use on level 2 under algae(1). See list F.

Chlorophyta
Includes use on level 2 under algae(1). See list F.

Chondrichthyes
Includes use on level 2 under Pisces(1). See list F.

Chondrostei
Term introduced in 1981. Includes use on level 2 under Pisces(1). See list F.

Chordata
Includes use on level 1 and 2 as a fossil term (list F). Term is to be used only when more specific terms do not apply.

chromates
Includes use on level 2 under minerals(1). See list L.

—— *CR see under* minerals

chromatography *CR see under* methods *under* chemical analysis

chromite ores
Term introduced in 1981. Includes use on level 1 and 2 as a commodity term (list C).

—— *CR see also under* economic geology *under* ——

chromium
Includes use on level 1 and 2 as a chemical element (list D).

Chrysophyta
Including golden-brown algae. Includes use on level 2 under algae(1). See list F.

circulation
A valid level 1 term through 1977 used in combination with oceans (i.e. oceans, circulation). After 1977, includes use under meteorology(1).

—— *see* climate-induced circulation; ocean circulation; thermal circulation; thermohaline circulation

circulation in the ocean *CR see* ocean circulation

Cirripedia
Includes use on level 2 under Arthropoda(1). See list F.

Cladocopina
Term introduced in 1981. Includes use on level 2 under Ostracoda(1). See list F.

classification
Includes use as level 2 or 3 term appropriate to a large number of topics, e.g. on level 2 under geochemistry(1) and under lava(1). See list G.

clastic rocks
Valid index term for GeoRef since 1976. After 1977, use clastic rocks on level 2 under sedimentary rocks(1).

—— *CR see under* sedimentary rocks

clastic sediments
Valid index term for GeoRef since 1976. Before 1978 included use in combination with terrigenous or nonterrigenous (i.e. clastic sediments, terrigenous) on level 2 under sediments(1). After 1978, use clastic sediments on level 2.

—— *CR see under* sediments

clay *CR see under* —— *under* sediments

clay mineralogy
Includes use on level 1 (list A). Term set options are:
areal studies
area
experimental studies
chemical property
material
mineral
physical property
subtopic
mineral data
mineral name

clay minerals
Includes use in combination with sheet silicates (i.e. sheet silicates, clay minerals) on level 2 under minerals(1). See list L.

clays
Includes use as a commodity term on level 1 (list C). Before 1981, included use when of economic value or of engineering application. As of 1981, use only when of economic value; use clay for material in engineering geology.

—— *CR see also* ceramic materials; kaolin; *see also under* economic geology *under* ——

cleavage *CR see under* —— *under* foliation

climate-induced circulation
Term introduced in 1978. Includes use on level 2 under ocean circulation(1).

clinoamphibole
As of 1981, includes use in combination with chain silicates (i.e. chain silicates, clinoamphibole) on level 2 under minerals(1). See list L. Before 1981, included use on level 3 under minerals(1) chain silicates, amphibole group(2).

clinopyroxene
As of 1981, includes use in combination with chain silicates (i.e. chain silicates, clinopyroxene) on level 2 under minerals(1). See list L. Before 1981, included use on level 3 under minerals(1) chain silicates, pyroxene group(2).

Clymeniida
Term introduced in 1981. Includes use on level 2 under Mollusca(1). See list F.

coal
Includes use on level 1 as a commodity term (list C); on level 3 under sedimentary rocks(1) organic residues(2).

—— CR see also under ____ under sedimentary rocks; see also under economic geology under ____
—— see brown coal

cobalt
Includes use on level 1 and 2 as a chemical element (list D). As of 1981, use cobalt ores for cobalt as a commodity. Before 1981, included use on level 1 as a commodity term.

cobalt ores
Term introduced in 1981. Includes use on level 1 and 2 as a commodity term (list C).
—— CR see also under economic geology under ____

Coccolithophoraceae
Includes use on level 2 under algae(1). See list F.

Codiaceae
As of 1981, includes use on level 2 under algae(1). See list F. Before 1981, included use on level 3 under algae(1) Chlorophyta(2).

Coelenterata
Includes use on level 1 and 2 as a fossil term (list F).

Coleopteroida
Term introduced in 1981. Includes use on level 2 under Insecta(1). See list F.

collections
Usage restricted to collections in museums, etc. Includes use on level 2 under meteorites(1); on level 3 under fossil names (list F), mineral names (list L), and paleontology(1). Also use collecting on level 3 under minerals(1).

Colombia
Includes use on level 1 as an area term (list O). For term set options see list B.

Colorado
Includes use on level 1 as an area term (list O). For term set options see list B.

Colorado Plateau
Index states as applicable. Introduced as level 1 area term in 1978.

colorimetry CR see under methods under chemical analysis

Columbia Plateau
A level 1 area term as of 1978. Lava basin. Index states as applicable.

columbium CR see niobium

comets
Includes use on level 2 under interplanetary space(1).

Commelinidae
Term introduced in 1981. Includes use on level 2 under angiosperms(1). See list F.

Comoro Islands
Group of volcanic islands in N Mozambique Channel between Mozambique and Malagasy Republic. An overseas territory of France. As of 1977 includes use as a level 1 area term (list O). See list B for term set options.

compensation
Includes use on level 2 under isostasy(1).

complexes see ring complexes

composition
Includes use as level 2 or 3 term appropriate to a large number of topics, e.g. on level 2 under meteorology(1). See list G.
—— see acidic composition; alkalic composition; calc-alkalic composition; calcic composition; ferromanganese composition; mafic composition; phosphate composition

compounds see organic compounds

concepts
Includes use as level 2 or 3 term appropriate to a large number of topics, e.g. on level 2 under geochemistry(1) and mineral exploration(1). See list G.

concretions CR see under secondary structures under sedimentary structures

conductivity see thermal conductivity

Condylarthra
Includes use on level 2 under Mammalia(1). See list F.

cones see shatter cones

Congo
This term is now limited to the present Congo (People's Republic of the Congo). Formerly it also included what is now Zaire. Includes use on level 1 as an area term (list O). For term set options see list B.

congresses CR see symposia

Coniferales
Includes use on level 2 under gymnosperms(1). See list F.

connate waters
Includes use on level 2 under ground water(1).
—— CR see under ground water

Connecticut
Includes use on level 1 as an area term (list O). For term set options see list B.

Conodonta
Term introduced in 1981. Includes use on level 1 and 2 as a fossil term (list F).

conodonts
As of 1981, restricted to use as the common name for Conodonta. See list F. To be used for Conodonta when they are used as tools in stratigraphy. Includes use on level 1. Before 1981, included use on level 1 and 2 as a fossil term. If 1, term set options are:

biostratigraphy
age [List E]

conservation
Level 1 term as of 1978. Includes use on level 2 under impact statements(1) and land use(1). Used for environmental conservation of natural resources. If 1, term set options are:
  environment
    beaches
    bogs
    caves
    ecosystems
    marshes
    reefs
    shorelines
  natural resources
    coal
    energy sources
    forests
    ground water
    metal ores
    mineral resources
    underground space
    water resouces
    wetlands
  topic [experimental studies, impact statements, methods, programs]
    subtopic
—— CR see also land use; reclamation; see also under environmental geology under ____

construction
As of 1978, term is used on level 2.

construction materials
Includes use on level 1 as a commodity term (list C).
—— CR see also dolostone; granite; gravel; limestone; marble; sand; sandstone; see also under economic geology under ____

contact metamorphism
Not a valid term from 1973 through 1977. After 1977, includes use on level 2 under metamorphism. As of 1981, also includes use on level 2 under intrusions(1).

contamination
As of 1981, restricted to use on level 2 under magmas(1) and is not to be confused with pollution. Before 1981, included use on level 2 under ground water(1).

continental drift
For general discussions of "classical" drift concepts as well as plate tectonic reconstructions. Includes use on level 1 (list A); on level 3 under plate tectonics(1). Term set options are:
  area [use terms like Gondwana, Iapetus, Laurasia, Mesogaea, Paleoafrica, Pangaea, Parathys, Tethys]
  locality

topic [causes, concepts, evolution, general, genesis, interpretation, mechanism, paleobotany, paleoclimatology, paleomagnetism, paleontology, patterns]
subtopic (no area term)

—— CR see also plate tectonics; sea-floor spreading; see also under stratigraphy under ____; see also under tectonophysics under ____

continental shelf
For studies of geological processes taking place on the shelf. Restricted to the modern shelf area. Includes use on level 1 (list A). Also used on level 3 in area sets under oceanography(2) where the level 1 term refers to the adjacent land area. A term for the ocean is also indexed for studies on specific continental shelves; however, the first level term is usually the land area. Term set options are:
topic [List B except for areal geology]
subtopic (no area term)

—— CR see also continental slope; marine geology; see also under oceanography under ____

continental slope
Includes use on level 1 (list A). For modern slope area. Term set options are:
topic [List B except for areal geology]
subtopic (no area term)

—— CR see also continental shelf; marine geology; see also under oceanography under ____

control see production control

controls
Includes use on level 2 under mineral deposits, genesis(1) and sedimentation(1).

Conularida
Term introduced in 1981. Includes use on level 2 under Coelenterata(1). See list F.

convection
Includes use on level 2 and 3 under meteorology(1) and under tectonophysics(1); on level 3 under plate tectonics(1).

convection currents
Includes use on level 2 under tectonophysics(1); on level 3 under Earth(1) and plate tectonics(1).

coordinate systems
Includes use on level 2 under magnetosphere(1).

coordinates see geodetic coordinates

coordination
As of 1978, term is used on level 2 under crystal chemistry(1).

Copepoda
Includes use on level 2 under Arthropoda(1). See list F.

copper
Includes use on level 1 and 2 as a chemical element (list D). As of 1981, use copper ores for copper as a commodity. Before 1981, included use on level 1 as a commodity term.

copper ores
Term introduced in 1981. Includes use on level 1 and 2 as a commodity term (list C).

—— CR see also under economic geology under ____

coprolites
Includes use on level 1 and 2 as a fossil term(list F); on level 3 under sedimentary structures(1) biogenic structures(2). See list K.

Coral Sea
Between Queensland, Australia, on the W and New Hebrides and New Caledonia on the E. As of 1977, includes use on level 1 as an area term (list O). For term set options see list B.

Corallimorpharia
Includes use on level 2 under Coelenterata(1). See list F.

Corallinaceae
As of 1981, includes use on level 2 under algae(1). Before 1981, included use on level 3 under algae(1) Rhodophyta(2). See list F.

corals
As of 1981, restricted to use as the common name for Coelenterata. See list F. To be used for Coelenterata when they are used as tools in stratigraphy. Includes use on level 1. Before 1981, included use on level 3 under Coelenterata(1) Anthozoa(2), and on level 3 as a reef constituent. If 1, term set options are:
biostratigraphy
age [List E]

—— CR see under ____ under Coelenterata; reefs

Cordaitales
Includes use on level 2 under gymnosperms(1). See list F.

core
Used for the core of the Earth. For that of other planets and Moon, see appropriate set. Includes use on level 1 (list A); on level 2 under Earth(1), and under seismology(1). Term set options are:
topic [composition, concepts, evolution, general, genesis, interpretation, processes, properties, temperature, theoretical studies]
subtopic

—— CR see also under seismology under ____; see also under tectonophysics under ____

Coriolis force
Includes use on level 2 under ocean circulation(1).

corona
Includes use on level 2 under astrophysics and solar physics(1).

correlation
Formerly a general term appropriate to a large number of topics (list G). As of 1976, the term was restricted to stratigraphical correlation. Includes use on level 2 under stratigraphy(1).

Corsica
Island in the Mediterranean Sea and a department of France. Includes use on level 1 as an area term (list O). For term set options see list B.

corundum deposits
Term introduced in 1981. Includes use on level 1 and 2 as a commodity term (list C). For corundum as a gem, index corundum and gems.

—— CR see also under economic geology under ____

Corynexochida
Includes use on level 2 under Trilobita(1). See list F.

cosmic dust
Includes use on level 2 under planetology(1) and (as of 1981) extraterrestrial geology(1).

cosmic rays
Includes use on level 2 under interplanetary space(1), magnetosphere(1), planetology(1), and (as of 1981) extraterrestrial geology(1); on level 3 under aeronomy(1) ionization(2).

cosmochemistry
For "geochemistry" of extraterrestrial materials. Includes use on level 2 under extraterrestrial geology(1), planetology(1), and asteroids(1) as of 1981. Before 1981, included use on level 3.

—— CR see under ____ under planetology

Costa Rica
Includes use on level 1 as an area term (list O). For term set options see list B.

Cotylosauria
Term introduced in 1981. Includes use on level 2 under Reptilia(1). See list F.

craters see meteor craters

creep
As of 1978, term is used on level 2 under permafrost(1) and slope stability(1).

Creodonta
Order. Includes on level 2 under Mammalia(1).

Cretaceous
Includes use on level 1 as an age term (list E); on level 2 under paleoterms, e.g. paleoecology, paleogeography, paleomagnetism. Above Jurassic, below Tertiary.
—— CR see also under geochronology under ——; see also under stratigraphy under ——
Crinoidea
Includes use on level 2 under Echinodermata(1). See list F.
Crocodilia
As of 1981, includes use on level 2 under Reptilia(1). See list F. Before 1981, included use on level 3 under Reptilia(1) Archosauria(2).
cross-bedding CR see under —— under sedimentary structures
crust
Used for the Earth's crust, and the crust of the planets and their satellites. As of 1981, for the crust of the Moon, use lunar crust. When used on level 1, or on level 2 or 3 in sets such as seismology(1) or plate tectonics(1), it refers to the crust of the Earth. When used in a planet set (when level 1 term is planet name), it refers to the crust of that planet. If 1, term set options are:
topic [age, anomalies, composition, concepts, evolution, genesis, interpretation, observations, processes, properties, theoretical studies, thickness]
subtopic (no area term)
—— CR see also under seismology under ——; see also under tectonophysics under ——
—— see lunar crust
Crustacea
Includes use on level 2 under Arthropoda(1). See list F.
cryptoexplosion features
Includes use on level 2 under geomorphology(1). Used for structures formed by the explosive release of energy, that exhibit intense rock deformation, and that show no obvious genetic relation to volcanism, tectonics, or meteorite impact.
—— CR see also under geomorphology
Cryptostomata
Includes use on level 2 under Bryozoa(1). See list F.
crystal chemistry
Used for the relations among chemical composition, structure and properties of crystals. Includes use on level 1 (list A) and on level 2 under minerals(1). Term set options are:
mineral group (List L)
mineral species
topic (if more than one single specific mineral; e.g. bonding, coordination, ion exchange, order-disorder, partitioning, phase equilibria)

subtopic
—— CR see also crystal growth; minerals
crystal growth
Used for studies on natural or artificial growth of crystals. Includes use on level 1 (list A). Term set options are:
mineral group (list L)
mineral species
topic (if more than one single specific mineral; e.g. crystal form, mechanism, phase equilibria, synthesis, twinning)
subtopic
—— CR see also crystal chemistry; minerals
crystal structure
Used for the internal structure of the crystal. Includes use on level 1 (list A) and level 2 under minerals(1). Term set options are:
mineral group (list L)
mineral species
topic (if more than one single specific mineral; e.g. bonding, cell dimensions, defects, refinement)
subtopic
—— CR see also crystal chemistry; minerals
crystallography
Treated as a whole. Used for general studies on the discipline of crystallography. Includes use on level 1 (list A). Term set options are:
topic [automatic data processing, bibliography, catalogs, classification, concepts, education, experimental studies, general, history, instruments, methods, nomenclature, objectives, observations, philosophy, practice, principles, symposia, textbooks, theoretical studies]
subtopic
—— CR see also mineralogy
Ctenodontida
Term introduced in 1981. Includes use on level 2 under Mollusca(1). See list F.
Ctenostomata
Includes use on level 2 under Bryozoa(1). See list F.
Cuba
Includes use on level 1 as an area term (list O). For term set options see list B.
curium
Includes use on level 1 and 2 as chemical element (list D).
current research
From 1978-1980, included use on level 2 under surveys(1). As of 1981, includes use on level 2 under survey organizations(1).

currents
Includes use on level 2 under ionosphere(1) and ocean circulation(1); on level 3 under aurora(1) electrical field(2) and magnetic field(2); on level 3 under meteorology(1) electrical phenomena(2).
—— see convection currents
Cyanophyta
Blue algae, blue-green algae and Schizophyta are included here. Includes use on level 2 under algae(1). See list F.
Cycadales
Includes use on level 2 under gymnosperms(1). See list F.
Cycadofilicales
Term introduced in 1981. Includes use on level 2 under gymnosperms(1). See list F.
cycles
Includes use on level 2 under paleoclimatology(1) and hydrology(1). Before 1982, also included use on level 2 under geochemistry(1). As of 1982, use geochemical cycle under geochemistry(1).
cyclic processes
Term introduced in 1978. Includes use on level 2 under sedimentation(1).
Cyclocystoidea
Includes use on level 2 under Echinodermata(1). See list F.
Cyclostomata
Includes use on level 2 under Bryozoa(1). See list F.
cylindrical structures
Includes use on level 2 under sedimentary structures(1). See list K.
Cynomorpha
Term introduced in 1981. Includes use on level 2 under Mammalia(1). See list F.
Cypridocopina
Term introduced in 1981. Includes use on level 2 under Ostracoda(1). See list F.
Cyprus
Island nation in the Mediterranean Sea. Includes use on level 1 as an area term (list O). For term set options see list B.
Cyrtodontida
Term introduced in 1981. Includes use on level 2 under Mollusca(1). See list F.
Cystoidea
Includes use on level 2 under Echinodermata(1). See list F.
Cytherocopina
Term introduced in 1981. Includes use on level 2 under Ostracoda(1). See list F.
Czechoslovakia
Includes use on level 1 as an area term (list O). For term set options see list B.

D-region
Includes use on level 2 under ionosphere(1).

dacites
Term introduced in 1981. Includes use on level 2 under igneous rocks(1). See list H.
—— *CR see under* igneous rocks

dams
Level 1 term as of 1978 used for geologic studies on dams. Includes use on level 2 under foundations(1). If level 1, term set options are:
　construction
　　subtopic
　design
　　subtopic
　foundations
　　subtopic
　site exploration
　　subtopic
—— *CR see also under* engineering geology *under* ——

Dasycladaceae
Includes use on level 2 under algae(1). See list F.

data *see* automatic data processing; mineral data

dates
Includes use on level 2 under absolute age(1).

dating *see* fission-track dating; isotope dating; particle-track dating; radioactive dating

debris flows
As of 1978, term is used on level 2 under slope stability(1).

defects
As of 1978, term is used on level 2 under crystal structure(1).

deformation
Used for the process of folding, faulting, shearing, compression, or extension of the rocks as a result of various forces. Only small-scale deformations considered here. See tectonics(1) for large scale. Includes use on level 1 (list A) and level 2 under rock mechanics(1) and soil mechanics(1). Term set options are:
　experimental studies
　　effects (e.g. creep, elastic strain, flow lines, fractures, kink-band structures, plastic flow, recrystallization, relaxation, twin-gliding) or name of modulus or limit (e.g. compression, elastic limit, fracture strength, shock metamorphism, tension, torsion, yield strength) or theory (e.g. elasticity, plasticity, viscoelasticity, viscosity) or material
　field studies
　　effects or name of modulus or limit or material
　theoretical studies

effects or name of modulus or limit or material
—— *CR see also* geophysics; structural analysis
—— *see* soft sediment deformation

Deinotherioidea
Term introduced in 1981. Includes use on level 2 under Mammalia(1). See list F.

Delaware
Includes use on level 1 as an area term (list O). For term set options see list B.

deltas *CR see under* environment *under* sedimentation; *see under* fluvial features *under* geomorphology

Demospongea
Includes use on level 2 under Porifera(1). See list F.

Dendroidea
Includes use on level 2 under Graptolithina(1). See list F.

Denmark
Includes use on level 1 as an area term (list O). For term set options see list B.

densities *see* ion densities and temperatures

densities and temperatures
Includes use on level 2 under aeronomy(1).

deposition
Includes use on level 2 under sedimentation(1); on level 3 under glacial geology(1) glaciation(2).

deposits *see* asbestos deposits; barite deposits; bentonite deposits; boron deposits; bromine deposits; calcite deposits; corundum deposits; dolostone deposits; evaporite deposits; feldspar deposits; genesis of ore deposits; glauconite deposits; granite deposits; graphite deposits; gravel deposits; gypsum deposits; heavy mineral deposits; iodine deposits; kaolin deposits; lead-zinc deposits; limestone deposits; magnesite deposits; marble deposits; mica deposits; mineral deposits; monazite deposits; nitrate deposits; nonmetal deposits; phosphate deposits; pumice deposits; rare earth deposits; sandstone deposits; slate deposits; sulfur deposits; talc deposits; vermiculite deposits; zircon deposits

depth
Includes use on level 2 under Mohorovicic discontinuity(1).

Dermapteroida
Term introduced in 1981. Includes use on level 2 under Insecta(1). See list F.

Dermoptera
Term introduced in 1981. Includes use on level 2 under Mammalia(1). See list F.

description *see* landform description

deserts *CR see under* —— *under* ecology; *see under* eolian features *under* geomorphology

design
As of 1978, includes use on level 2 under dams(1), foundations(1), marine installations(1), nuclear facilities(1), reservoirs(1), shorelines(1), tunnels(1), underground installations(1), and waterways(1).

Desmidiales
Term introduced in 1981. Includes use on level 2 under algae(1). See list F.

Desmoceratida
Term introduced in 1981. Includes use on level 2 under Mollusca(1). See list F.

Desmostylia
Includes use on level 2 under Mammalia(1). See list F.

detection
Includes use as a level 2 or 3 term appropriate to a large number of topics, e.g. on level 2 under earthquakes(1). See list G.

deuterium
Includes use on level 1 and 2. See list D (chemical elements).
—— *CR see also* hydrogen; tritium

Devonian
Includes use on level 1 as an age term (list E); on level 2 under paleoterms, e.g. paleoecology, paleogeography, paleomagnetism. Above Silurian, below Carboniferous.
—— *CR see also under* geochronology *under* ——; *see also under* stratigraphy *under* ——

diabase
As of 1981, includes use on level 2 under igneous rocks(1). See list H. Before 1981, included use on level 3 under igneous rocks(1) basalt family(2).
—— *CR see under* igneous rocks

diagenesis
Level 1 term as of 1978, used for all the chemical, physical and biological changes, modifications, or transformations undergone by a sediment after its initial deposition, and during and after its lithification, exclusive of weathering and metamorphism. Includes use on level 2 under sedimentary rocks(1) and sediments(1). If level 1, term set options are:
　controls
　　subtopic
　dolomitization
　　subtopic
　effects
　　subtopic
　geochemistry

subtopic indicators
subtopic materials
subtopic mechanism
subtopic processes
subtopic sedimentation
subtopic topic [e.g. environment, experimental studies, models, theoretical studies]
subtopic
—— CR see also sedimentation
diamond CR see under native elements under minerals
diamonds
Use only when of economic value. Includes use on level 1 as a commodity term (list C);on level 2 under placers(1).
—— CR see also under economic geology under ——
diapirs CR see under —— under folds; salt tectonics
diastrophism CR see epeirogeny; orogeny; tectonics
diatomite
Includes use on level 1 as a commodity term (list C). As of 1981, to be used only if of economic value; otherwise, use diatomaceous earth. Before 1981, included use on level 3 under sedimentary rocks(1) clastic rocks(2).
—— CR see also under —— under sedimentary rocks; see also under economic geology under ——
diatoms
Includes use on level 2 under algae(1). See list F.
—— CR see under algae
diatremes
Includes use on level 2 under intrusions(1).
—— CR see under intrusions
Dibranchiata
Term introduced in 1981. Includes use on level 2 under Mollusca(1). See list F.
Dicotyledoneae
Includes use on level 2 under angiosperms(1). See list F.
dictionaries CR see glossaries; lexicons
Dictyonellidina
Term introduced in 1981. Includes use on level 2 under Brachiopoda(1). See list F.
Didymograptina
Includes use on level 2 under Graptolithina(1). See list F.
differential thermal analysis
Used on level 1 for methodology through 1977. After 1977, includes use on level 2 under thermal analysis(1).

—— CR see under thermal analysis
differentiation
Includes use on level 2 under magmas(1) and on level 3 under intrusions(1), igneous rocks(1), and Earth(1).
diffraction
Includes use on level 3 under seismology(1).
diffusion
Includes use on level 2 under aeronomy(1), meteorology(1), and ocean circulation(1); on level 3 under geochemistry(1) processes(2). See also under specific elements. Included use on level 2 under oceanography(1) until 1976.
dikes
Includes use on level 2 under intrusions(1).
—— CR see under intrusions
Dilleniidae
Term introduced in 1981. Includes use on level 2 under angiosperms(1). See list F.
Dinocerata
Term introduced in 1981. Includes use on level 2 under Mammalia(1). See list F.
Dinoflagellata
Hystrichosphaeridae is included here and under acritarchs. Includes use on level 2 under palynomorphs(1). See list F.
—— CR see under palynomorphs
dinosaurs
As of 1981, includes use on level 2 under Reptilia(1). See list F. Before 1981, included use on level 3 under Reptilia(1).
—— CR see Archosauria under Reptilia
diorite CR see under —— under igneous rocks
diorites
Term introduced in 1981. Includes use on level 2 under igneous rocks(1). See list H.
—— CR see under igneous rocks
—— see alkali diorites; quartz diorites
Diplograptina
Includes use on level 2 under Graptolithina(1). See list F.
dipmeter logging
Term introduced in 1978. Includes use on level 2 under well-logging(1).
Dipnoi
Term introduced in 1981. Includes use on level 2 under Pisces(1). See list F.
Diprotodonta
Term introduced in 1981. Includes use on level 2 under Mammalia(1). See list F.

Discorbacea
Includes use on level 2 under foraminifera(1). See list F.
displacements
Includes use on level 2 under faults(1).
disposal see waste disposal
distribution
Includes use as a level 2 or 3 term appropriate to a large number of topics, e.g. on level 2 under faults(1), folds(1), fractures(1) and under ocean circulation(1). See list G.
District of Columbia
Includes use on level 1 as an area term (list O). For term set options see list B.
disturbances
Term is restricted to extraterrestrial phenomena. Includes use on level 2 under ionosphere(1).
Djibouti
Became independent June 27, 1977. Formerly French Territory of the Afars and Issas. Capital city also named Djibouti. As of 1978, includes use on level 1 as an area term.
Docodonta
Includes use on level 2 under Mammalia(1). See list F.
dolomite CR see under —— under ——
dolomitization
As of 1978, term is used on level 2 under diagenesis(1).
—— CR see under —— under metasomatism; sedimentation; see under diagenesis
dolostone CR see also under —— under sedimentary rocks
dolostone deposits
Term introduced in 1981. Includes use on level 1 and 2 as a commodity term (list C).
—— CR see also under economic geology under ——
domains see magnetic domains
domes CR see under —— under folds
—— see salt domes
Dominican Republic
Occupies eastern two thirds of island of Hispaniola in the Greater Antilles. Includes use on level 1 as an area term (list O). For term set options see list B.
drift see continental drift
drumlins CR see under glacial features under glacial geology
dunes CR see under eolian features under geomorphology
dust see cosmic dust; interplanetary dust

dynamic metamorphism
Term introduced in 1978. Includes use on level 2 under metamorphism(1).

dynamics
As of 1978, term is used on level 2 under shorelines(1).

dysprosium
Includes use on level 1 and 2 as chemical element (list D).

E-region
Includes use on level 2 under ionosphere(1).

Earth
Treated as a whole. For general concepts or studies. Includes use on level 1(list A). Term set options are:
topic [age, composition, concepts, evolution, general, genesis, gravity field, interior, magnetic field, motions, observations, processes, properties, temperature, theoretical studies]
subtopic [e.g. Chandler wobble, convection currents, density, differentiation, Earth tides, expansion, free oscillations, melting, rotation]
—— see figure of Earth; rare earth deposits

earth pressure
As of 1978, term is used on level 2 under soil mechanics(1).

Earth tides CR see under processes under Earth; see under surveys under geodesy

Earth-current methods
Includes use on level 2 under geophysical methods(1).

Earth-current surveys CR see under geophysical surveys under ——

earthflows
As of 1978, term is used on level 2 under slope stability(1).

earthquakes
Used on level 1 and 2 for studies emphasizing individual earthquakes. Includes use on level 2 under geologic hazards(1), mantle(1) and seismology(1). If level 1, term set options are:
topic [aftershocks, causes, classification, detection, effects, epicenters, focal mechanism, focus, genesis, magnitude, prediction, seismic intensity]
subtopic [e.g. deep-focus earthquakes, intermediate-focus earthquakes, shallow-focus earthquakes, or year, for specific earthquake] (no area term)
—— CR see also engineering geology; seismology; see also under engineering geology under ——; see also under seismology under ——

earths see rare earths

earthworks
As of 1978, term is used on level 2 under foundations(1).
—— CR see under foundations

East China Sea
Enclosed by China, Korea, Japan, and Taiwan. As of 1977, includes use on level 1 as an area term (list O).

East Germany
Officially German Democratic Republic or Deutsche Demokratische Republik. In N central Europe, bounded on N by the Baltic Sea, on E by Poland, on S by Czechoslovakia and West Germany, and on W by West Germany. Introduced as a level 1 area term in 1978.

East Pacific Ocean Islands
Term introduced in 1981. Includes use on level 1 as an area term (list O). For term set options see list B.

Eastern Hemisphere
Used when discussing many large areas too numerous to mention. Includes use on level 1 and 2 as an area term (list O). If 1, see term set options under list B.
—— CR see also Africa; Antarctic Ocean; Antarctica; Arctic Ocean; Asia; Atlantic Ocean; Eurasia; Europe; Indian Ocean; USSR

Eastern U.S.
Term introduced in 1978. Includes use as a level 1 area term.

ebridians
Includes use on level 2 under Protista(1). See list F.

Echinodermata
Includes use on level 1 and 2 as a fossil term (list F).

echinoderms
Term introduced in 1981 as the common name for Echinodermata. See list F. To be used for Echinodermata when they are used as tools in stratigraphy. Includes use on level 1. If 1, term set options are:
biostratigraphy
age [List E]

Echinoidea
Includes use on level 2 under Echinodermata(1). See list F.

Echiurida
Term introduced in 1981. Includes use on level 2 under worms(1). See list F.

eclogite
As of 1981, includes use on level 2 under metamorphic rocks(1). See list J. Before 1981, included use on level 3 under metamorphic rocks(1) granulites(2).

ecology
Includes use on level 1 (list A); on level 2 under fossil groups (list F). Used for the study of relationships between organisms and their environments, including the study of communities, patterns of life, natural cycles, relationships of organisms to each other, biogeography, and population changes. If 1, term set options are:
name of fossil group (List F) (for Recent fauna and flora)
type of environment [e.g. arctic environment, alpine environment, boreal environment, brackish-water environment, coastal environment, deltaic environment, eolian environment, estuarine environment, glacial environment, intertidal environment, lacustrine environment, lagoonal environment, marine environment, paludal environment, reefs, terrestrial environment]
topic [analysis, changes, concepts, evolution, interpretation, observations, processes]
subtopic (no area term)
—— CR see also under environmental geology under ——; see also under paleobotany under ——; see also under paleontology under ——
—— see human ecology

economic geology
For the discipline as a whole. See also specific commodities (list C). Includes use on level 1 (list A); on level 2 under area terms (list B). If 1, term set options are:
topic [applications, bibliography, catalogs, classification, concepts, education, experimental studies, general, history, instruments, interpretation, methods, nomenclature, objectives, philosophy, practice, principles, symposia, textbooks, theoretical studies]
subtopic

economics
Includes use on level 2 under commodity terms (list C) for general treatments not tied to specific areas.

ecosystems CR see under —— under conservation; ecology

Ectotropha
Includes use on level 2 under Insecta(1). See list F.

Ecuador
Includes use on level 1 as an area term (list O). For term set options see list B.

Edentata
Includes use on level 2 under Mammalia(1). See list F.

Edrioasteroidea
Includes use on level 2 under Echinodermata(1). See list F.

Edrioblastoidea
Includes use on level 2 under Echinodermata(1). See list F.

education
Includes use on level 1 (list A) for general discussions of methodology and curricula in education; on level 2 under major disciplines; as a level 2 or 3 term appropriate to a large number of topics (list G). If 1, term set options are:
topic [List B (except for areal geology), or extraterrestrial geology, general, geophysics]
subtopic (college-level education, curricula, elementary geology, elementary school, graduate-level education, high school, junior high school, materials, methods, objectives, popular geology, vocational school)

effects
Includes use as a level 2 or 3 term appropriate to a large number of topics, e.g. on level 2 under earthquakes(1) and under faults(1). See list G.

—— *see* geomorphologic effects

Egypt
Includes use on level 1 as an area term (list O). For term set options see list B.

Ekman spiral
Includes use on level 2 under ocean circulation(1).

El Salvador
Includes use on level 1 as an area term (list O). For term set options see list B.

Elasmobranchii
As of 1981, includes use on level 2 under Pisces(1). Before 1981, included use on level 3 under Pisces(1) Chondrichthyes(2). See list F.

elastic waves
Do not use seismic waves. Includes use on level 2 under seismology(1) and under mantle (1); on level 3 under earthquakes(1).

—— *CR see under* explosions; seismology

elasticity
As of 1978, term is used on level 2 under rock mechanics(1) and soil mechanics(1).

—— *CR see under* ____ *under* deformation; *see under* rock mechanics; soil mechanics

electrical field
Includes use on level 2 under aurora(1), ionosphere(1), and magnetosphere(1).

electrical logging
Not a valid term from 1971 through 1977. Now includes use on level 2 under well-logging(1).

—— *CR see* well-logging

electrical methods
Includes use on level 2 under geophysical methods(1).

electrical phenomena
Includes use on level 2 under meteorology(1).

electrical surveys *CR see under* geophysical surveys *under* ____

electrojet
Includes use on level 2 under ionosphere(1); on level 3 under aurora(1) electrical field(2) and magnetic field(2).

electromagnetic logging
Term introduced in 1978. Includes use on level 2 under well-logging(1).

electromagnetic methods
Includes use on level 2 under geophysical methods(1).

electromagnetic radiation
Includes use on level 2 under astrophysics and solar physics(1) and under interplanetary space(1).

electromagnetic surveys *CR see under* geophysical surveys *under*

electromagnetic waves
Includes use on level 2 under meteorology(1).

electron content
Includes use on level 2 under ionosphere(1).

electron microscopy
Includes use on level 1 (list A) for methodology, not for data. If 1, term set options are:
applications
subtopic
instruments
subtopic
methods
subtopic
techniques
subtopic

electron probe *CR see under* methods *under* chemical analysis; spectroscopy

elements *see* migration of elements; native elements; trace elements

Elephantoidea
Term introduced in 1981. Includes use on level 2 under Mammalia(1). See list F.

embankments
As of 1978, term is used on level 2 under highways(1) and slope stability(1).

—— *CR see under* highways; slope stability

Embrithopoda
Includes use on level 2 under Mammalia(1). See list F.

emplacement
Includes use on level 2 under intrusions(1).

energy sources
Includes use on level 1 as a commodity term (list C). See also names of appropriate commodities.

—— *CR see also* fuel resources; geothermal energy; natural gas; petroleum; uranium ores; *see also under* economic geology *under* ____

engineering *see* petroleum engineering

engineering geology
Used for geology as applied to engineering practice, especially mining and civil engineering. See appropriate features and processes under geomorphology(1). Includes use on level 1 (list A); on level 2 under area terms (list B). If 1, term set options are:
field studies
subtopic
materials, properties
(for material not covered under rock mechanics and soil mechanics)
material or property (e.g. elastic strain)
methods
type of method (cartography, photogeology, photogrammetry)
topic [bibliography, education, experimental studies, feasibility studies, frost action, maps, petroleum engineering, practice, seepage, site exploration, symposia, techniques]
subtopic

—— *CR see also* dams; deformation; earthquakes; environmental geology; foundations; geodesy; geologic hazards; geophysical methods; ground water; highways; impact statements; land subsidence; land use; marine installations; mining geology; nuclear facilities; reservoirs; rock mechanics; shorelines; slope stability; soil mechanics; tunnels; underground installations; waste disposal; waterways

engineering properties
Includes use on level 3 under soils(1).

England
Includes use on level 1 as an area term (list O). For term set options see list B.

English Channel
Strait between England and France connecting Atlantic Ocean with the North Sea. Includes use on level 1 as an area term (list O). For term set options see list B.

English Channel Islands
Term introduced in 1981 for the British group of islands in the English Channel off the W coast of Normandy. Before 1981, "Channel Is-

lands" was used for both these and the Santa Barbara Islands off California. Includes use on level 1 as an area term (list O). For term set options see list B.

Enteropneusta
Term introduced in 1981. Includes use on level 2 under Hemichordata(1). See list F.

Entomotaeniata
Term introduced in 1981. Includes use on level 2 under Mollusca(1). See list F.

Entomozocopina
Term introduced in 1981. Includes use on level 2 under Ostracoda(1). See list F.

Entotropha
Term introduced in 1981. Includes use on level 2 under Insecta(1). See list F.

environment
Includes use as a level 2 or 3 term appropriate to a large number of topics, e.g. on level 2 under sedimentation(1). See list G.

—— *see* arid environment

environmental analysis
For papers discussing the paleoenvironmental implications of several rock types and for papers whose main purpose is to determine environment. Includes use on level 2 under sedimentary rocks(1), sediments(1), and sedimentary structures(1).

environmental geology
Includes use on level 1 (list A); on level 2 under area terms (list B). Used for studies involving the collection, analysis, and application of geologic data and principles to problems created by human occupancy and use of the physical environment. If 1, term set options are:
topic [concepts, education, maps, symposia]
subtopic [controls, floods, human ecology, regional planning, urban planning]

—— *CR see also* conservation; ecology; engineering geology; geologic hazards; impact statements; land use; pollution; reclamation; waste disposal

Eocene
World. Above Paleocene, below Oligocene. Includes use on level 1 as an age term (list E); on level 2 under paleo- terms, e.g. paleoecology, paleogeography, paleomagnetism.

—— *CR see also under* geochronology *under* _____; *see also under* stratigraphy *under* _____

eolian features
Includes use on level 2 under geomorphology(1).

—— *CR see under* geomorphology

Eosuchia
Term introduced in 1981. Includes use on level 2 under Reptilia(1). See list F.

epeirogeny
Used for specific epeirogenic activity. For general treatment, see tectonics. Includes use on level 1 (list A). Term set options are:
age [list E]
area

—— *CR see also* neotectonics; orogeny; tectonics

Ephedrales
Term introduced in 1981. Includes use on level 2 under gymnosperms(1). See list F.

Ephemeropteroida
Term introduced in 1981. Includes use on level 2 under Insecta(1). See list F.

epicenters
Includes use on level 2 under earthquakes(1).

—— *CR see under* earthquakes

epidote group
Includes use in combination with orthosilicates (i.e. orthosilicates, epidote group) on level 2 under minerals(1). See list L.

Equatorial Guinea
Formerly Spanish Guinea. Comprises province of Rio Muni on the mainland and the province of Fernando Po consisting of the islands of Fernando Po and Annobon. Includes use on level 1 as an area term (list O). For term set options see list B.

equilibria *see* phase equilibria

erbium
Includes use on level 1 and 2 as a chemical element (list D).

Eridostraca
Term introduced in 1981. Includes use on level 2 under Ostracoda(1). See list F.

erosion *CR see under* slope stability; *see under* processes *under* geomorphology

—— *see* soil erosion

erosion features
Includes use on level 2 under geomorphology(1).

erosion surfaces *CR see under* erosion features *under* geomorphology

eskers *CR see under* glacial features *under* glacial geology

estuaries *CR see under* environment *under* sedimentation; *see under* fluvial features *under* geomorphology

Ethiopia
Includes use on level 1 as an area term (list O). For term set options see list B.

Eubradyodonti
Term introduced in 1981. Includes use on level 2 under Pisces(1). See list F.

Eurasia
Land mass comprising the continents of Europe and Asia. Index continents as applicable. Includes use on level 1 or 2 as an area term (list O). If 1, see term set options under list B.

—— *CR see also* Black Sea; Caspian Sea

Europe
Includes use on level 1 or 2 as an area term (list O). If 1, see term set options under list B.

—— *CR see also* the individual nations

europium
Includes use on level 1 and 2 as a chemical element (list D).

Euryapsida
Includes use on level 2 under Reptilia(1). See list F.

evaluation
Includes use on level 2 under mining geology(1).

evaporite deposits
Term introduced in 1981. Includes use on level 1 and 2 as a commodity term (list C).

evaporites *CR see also* gypsum; nitrates; potash; salt; sodium carbonate; sodium sulfate; *see also under* _____ *under* sedimentary rocks; *see also under* economic geology *under* _____

evolution
Includes use as a level 2 or 3 term appropriate to a large number of topics, e.g. on level 2 under paleontology(1), tectonics(1) and under fossil groups (list F). See list G (general terms).

—— *see* landform evolution

excavations
As of 1978, includes use on level 2 under tunnels(1), rock mechanics(1), and underground installations(1). Used under engineering geology(1) through 1977.

experimental studies
Includes use as a level 2 or 3 term appropriate to a large number of topics, e.g. on level 2 under meteorology(1), folds(1), fractures(1), heat flow(1), and under oceanography(1). See list G.

exploration
Includes use on level 2 or 3 under commodity terms(1), e.g. on level 2 under energy sources(1) and placers(1). See list C.

—— *see* mineral exploration; site exploration

explosions
A level 1 term as of 1978. Includes use on level 2 under geologic hazards(1) and seismology(1). Used for geotechnical and seismological studies on the effects of explosions. If 1, term set options are:
applications
  subtopic
chemical explosions
  subtopic
detection
  subtopic
effects
  subtopic
elastic waves
  subtopic
excavations
  subtopic
experimental studies
  subtopic
materials, properties
  subtopic
nuclear explosions
  subtopic
site exploration
  subtopic
theoretical studies
  subtopic
—— *CR see also under* engineering geology *under* ——; *see also under* seismology *under* ——
—— *see* chemical explosions; nuclear explosions

exposure age
Includes use on level 2 under geochronology(1).
—— *CR see under* geochronology

extraterrestrial geology
Broad topic for studies about materials and processes outside the Earth. Includes use on level 2 under automatic data processing(1), associations(1), and symposia(1). When used on level 1 (beginning in 1981), the universe as a whole may be implied. If 1, term set options are:
theoretical studies
  subtopic
topic [concepts, cosmic dust, cosmic rays, cosmochemistry, evolution, observations]
  subtopic

F-region
Includes use on level 2 under ionosphere(1).

facies
Includes use on level 2 under metamorphic rocks(1).

Faeroe Islands
Self governing Danish island group between Iceland and the Shetland Islands. Includes use on level 1 as an area term (list O). For term set options see list B.

failures
Term introduced in 1981. Includes use on level 2 under rock mechanics(1) and slope stability(1).

Falkland Islands
British colony 300 miles E of Straits of Magellan. Islands are claimed by Argentina. Includes use on level 1 as an area term (list O). For term set options see list B.

Far East
Easternmost Asia along the Pacific Ocean. Index countries and regions as applicable. Includes use on level 1 as an area term (list O). For term set options see list B.

fatty acids
Includes use on level 2 under organic materials(1).

faults
To be used for studies primarily stressing individual faults or systems of faults. For relationships with other structures, see tectonics. Includes use on level 1 (list A); on level 2 under structural analysis(1), geologic hazards(1), and nuclear facilities(1). If level 1, term set options are:
displacements
  subtopic [diagonal-slip faults, dip-slip faults, gravity faults, normal faults, overthrust faults, reverse faults, strike-slip faults, thrust faults, transcurrent faults, transform faults, wrench faults]
distribution
  subtopic (e.g. topic or area)
effects
  subtopic [breccia, gouge, mullions, mylonites, shear zones, slickensides, ultramylonite]
mechanics
  subtopic [e.g. compression, flexure, shear, stick-slip]
orientation
  subtopic [arcuate faults, bedding faults, dip faults, longitudinal orientation, oblique orientation, strike faults, transverse faults]
patterns
  subtopic [en echelon faults, parallel faults, peripheral faults, radial faults]
systems
  subtopic [block structures, grabens, horsts, rift zones, step faults]
theoretical studies
  subtopic

faunal studies
Includes use on level 2 under fossil group(1) for many classes or orders, or for genera. See list F.

feasibility studies
As of 1978, term is used on level 2.

features *see* bottom features; cryptoexplosion features; eolian fea-

tures; erosion features; fluvial features; glacial features; impact features; lacustrine features; periglacial features; shore features; solution features; volcanic features

feldspar *CR see also under* —— *under* minerals
—— *see* alkali feldspar; barium feldspar

feldspar deposits
Term introduced in 1981. Includes use on level 1 and 2 as a commodity term (list C).
—— *CR see also under* economic geology *under* ——

feldspar group
Includes use in combination with framework silicates (i.e. framework silicates, feldspar group) on level 2 under minerals(1). See list L.

ferns
As of 1981, includes use on level 1. See list F. To be used for pteridophytes when they are used as tools in stratigraphy. Before 1981, included use on level 3. If 1, term set options are:
biostratigraphy
age [List E]

ferromanganese composition
Term introduced in 1978. Includes use on level 2 under nodules(1).

fertilizers
Includes use on level 2 under soils(1).

field *see* electrical field; gravity field; magnetic field

field studies
Includes use on level 2 under deformation(1), and soils(1).

fields *see* oil and gas fields

figure of Earth
Includes use on level 2 under geodesy(1).

Fiji
Independent state consisting of island group between New Caledonia and Samoa in S Pacific Ocean. Includes use on level 1 as an area term (list O). For term set options see list B.

Filicopsida
Including Filicales. Includes use on level 2 under pteridophytes(1). See list F.

Finland
Includes use on level 1 as an area term (list O). For term set options see list B.

firn *CR see under* periglacial features *under* glacial geology

fish
As of 1981, includes use on level 1. See list F. To be used for Pisces when they are used as tools in stratigraphy. Before 1981, included use on level 3. If 1, term set options are:

biostratigraphy
age [List E]
fission-track dating
Term introduced in 1978. Includes use on level 2 under geochronology(1).
—— *CR see under* geochronology
Fissipeda
Term introduced in 1981. Includes use on level 2 under Mammalia(1). See list F.
fjords *CR see under* shore features *under* geomorphology
flares *see* solar flares
floods
As of 1978, term is used on level 2 under geologic hazards(1) and waterways(1).
—— *CR see under* ____ under hydrology; *see under* geologic hazards
floors *see* ocean floors
floral studies
Includes use on level 2 under fossil group(1) for many classes or orders, or for genera. See list F.
Florida
Includes use on level 1 as an area term (list O). For term set options see list B.
flow *see* heat flow
flow mechanism
Includes use on level 2 under lava(1).
flows *see* debris flows
fluctuations *see* sea-level fluctuations
fluid inclusions
A level 1 term as of 1978. Used for gaseous or liquid inclusions found in crystals. Before 1978, included use on level 2 under inclusions(1). If 1, term set options are:
topic [analysis, changes, composition, detection, experimental studies, genesis, geochemistry, geologic barometry, geologic thermometry, interpretation, P-T conditions, paleosalinity, temperature, theoretical studies]
type of inclusion [e.g. ammonia, carbon dioxide, gases, methane] or subtopic
—— *CR see also* inclusions
fluoborates
As of 1981, includes use on level 2 under minerals(1). See list L. Before 1981, included use on level 3 under minerals(1) halides(2).
—— *CR see under* minerals
fluorides
As of 1981, includes use on level 2 under minerals(1). See list L. Before 1981, included use on level 3 under minerals(1) halides(2).

fluorine
Includes use on level 1 and 2 as a chemical element (list D).
fluorspar
Use only when of economic value; otherwise use fluorite. Includes use on level 1 and 2 as a commodity term (list C).
fluosilicates
As of 1981, includes use on level 2 under minerals(1). See list L. Before 1981, included use on level 3 under minerals(1) halides(2).
—— *CR see under* minerals
fluvial features
Includes use on level 2 under geomorphology(1).
flux *see* meteorite flux
focal mechanism
As of 1981, includes use on level 2 under earthquakes(1). Before 1981, included use on level 3.
focus
Includes use on level 2 under earthquakes(1).
folds
Used for studies on folds and not those on several types of structure. Includes use on level 1 (list A); on level 2 and 3 under structural analysis(1). If 1, term set options are:
distribution
subtopic
experimental studies
subtopic
geometry
subtopic [cylindrical folds, plane cylindrical folds, plane noncylindrical folds]
mechanics
subtopic [compaction, decollement, flexural-slip, flexure, kinkband structures, shear]
orientation (attitude of fold elements with respect to external coordinates)
a. (folds defined on the basis of orientation in relation to the geographic horizontal plane): [horizontal orientation, inclined folds, nappes, normal folds, overturned folds, plunging folds, recumbent folds, vertical orientation]
b. (orientation of folds relative to spatially associated macroscopic structures such as large folds, fold systems, and orogenic zones): [discordant folds, drag folds, longitudinal orientation, oblique orientation, superposed folds]
style
subtopic [anticlines, antiform folds, asymmetric folds, basins, chevron folds, cleavage folds, concentric folds, conjugate folds, convolute folds, cross folds, diapirs, disharmonic folds, domes,

harmonic folds, intrafolial folds, isoclinal folds, kink folds, monoclines, polyclinal folds, ptygmatic folds, similar folds, symmetric folds, synclines, synform folds, troughs]
systems
subtopic [anticlinoria, en echelon folds, synclinoria]
theoretical studies
subtopic
foliation
Includes use on level 1 (list A); on level 2 and 3 under structural analysis(1). Used for planar arrangements of textural or structural features in any type of rock. If 1, term set options are:
genesis
subtopic [flow cleavage, shear cleavage]
interpretation
subtopic
style
subtopic [axial-plane structures, cleavage, fracture cleavage, lamination, schistosity, slaty cleavage, slip cleavage]
—— *CR see also* structural analysis
foraminifera
Includes use on level 1 and 2 as a fossil term (list F).
foraminifers
Term introduced in 1981 as the common name for foraminifera. See list F. To be used for foraminifera when they are used as tools in stratigraphy. Includes use on level 1. If 1, term set options are:
biostratigraphy
age [List E]
forests *CR see under* ____ under conservation
Formosa *CR see* Taiwan
fossil man
Term introduced in 1978. Includes use on level 1 and 2 as a fossil term (list F). Used mainly for archaeological and anthropological studies, but as of 1981, is restricted to papers about actual fossil remains. If taxonomy is given, use Mammalia.
fossil wood *CR see under* ____ *under* gymnosperms; Plantae
fossilization
Includes taphonomy. Includes use on level 2 under fossil group(1); on level 2 under paleontology(1). See list F.
fossils *CR see* appropriate fossil group
fossils, problematic *CR see* problematic fossils
foundations
A level 1 term as of 1978. Includes use on level 2 under dams(1). Used for geological studies on engineering foundations. If 1, term set options are:

bridges
subtopic
buildings
subtopic
construction
subtopic
dams
subtopic
design
subtopic
earthworks
subtopic
experimental studies
subtopic
highways
subtopic
instruments
subtopic
land subsidence
subtopic
materials, properties
subtopic
piles
subtopic
seepage
subtopic
settlement
subtopic
site exploration
subtopic
stability
subtopic
structures
subtopic
theoretical studies
subtopic
—— *CR see also* rock mechanics; soil mechanics; *see also under* engineering geology *under* ——
fractionation
Includes use on level 2 under isotopes(1); on level 3 under geochemistry(1) processes(2). Use fractional crystallization for fractionation of magmas.
fractures
Includes use on level 1 (list A); on level 3 under deformation(1). Used for a break in a rock usually without displacement. If 1, term set options are:
distribution
subtopic
experimental studies
subtopic
genesis
subtopic [release, shear, tension]
patterns
subtopic
style
subtopic [bedding, closed fractures, columnar joints, concentric fractures, conical fractures, cross fractures, dip fractures, extension fractures, feather fractures, joints, latent fractures, oblique fractures, polygonal fractures, radiating fractures, ring sheeting, strike fractures]

systems
subtopic
theoretical studies
subtopic
framework silicates
Includes use on level 2 under minerals(1); in combination with feldspar group, alkali feldspar, plagioclase, barium feldspar, nepheline group, scapolite group, silica minerals, sodalite group, and zeolite group (i.e. framework silicates, feldspar group) to form terms on level 2 under minerals(1). See list L.
—— *CR see under* minerals
France
Includes use on level 1 as an area term (list O). For term set options see list B.
French Guiana
French overseas department on NE coast. Includes use on level 1 as an area term (list O). For term set options see list B.
frost action
Includes use on level 2 under engineering geology(1) and geomorphology(1).
—— *CR see under* engineering geology; geomorphology; permafrost; rock mechanics; soil mechanics
frost heaving
As of 1978, term is used on level 2 under permafrost(1).
fuel resources
Term introduced in 1981 as a grouper for petroleum, oil shale, oil sands, and natural gas. Includes use on level 1 and 2 as a commodity term (list C).
—— *CR see also under* economic geology *under* ——
fumaroles
As of 1978, term is used on level 2 under thermal waters(1).
—— *CR see under* —— *under* thermal waters; volcanology
fungi
Includes use on level 1 and 2 as a fossil term (list F); on level 3 under soils(1) biota(2).
Fusulinidae
Includes use on level 2 under foraminifera(1). See list F.
Fusulinina
Includes use on level 2 under foraminifera(1). See list F.
gabbro *CR see under* —— *under* igneous rocks
gabbros
Term introduced in 1981. Includes use on level 2 under igneous rocks(1). See list H.
—— *CR see under* igneous rocks
—— *see* alkali gabbros

Gabon
Formerly part of French Equatorial Africa. Includes use on level 1 as an area term (list O). For term set options see list B.
gadolinium
Includes use on level 1 and 2 as a chemical element (list D).
Galapagos Islands
A territory of Ecuador about 600 miles W of mainland. Includes use on level 1 as an area term (list O). For term set options see list B.
gallium
Includes use on level 1 and 2 as a chemical element (list D).
Gambia
Includes use on level 1 as an area term (list O). For term set options see list B.
garnet *CR see under* —— *under* ——
garnet group
Includes use in combination with orthosilicates (i.e. orthosilicates, garnet group) on level 2 under minerals(1). See list L.
garnetite
Term introduced in 1981. Includes use on level 2 under metamorphic rocks(1). See list J.
gas *see* helium gas; natural gas; oil and gas fields
gas, natural *CR see* natural gas
gases *see* inert gases; noble gases
Gastropoda
Includes use on level 2 under Mollusca(1). See list F.
gems
Includes use on level 1 as a commodity term (list C). See also specific gems.
—— *CR see also under* economic geology *under* ——
general
Includes use as a level 2 or 3 term appropriate to a large number of topics, e.g. on level 2 under Mohorovicic discontinuity(1). As of 1976, the term is to be used on level 2 only when referring to geology. See list G.
genesis
Includes use on level 1 in combination with mineral deposits (i.e. mineral deposits, genesis); as a level 2 or 3 term appropriate to a large number of topics, e.g. on level 2 under foliation(1), fractures(1), and organic materials(1). See list C (commodities) and list G (general terms). As of 1981, also includes use on level 2 under ocean waves(1).
genesis of ore deposits *CR see* mineral deposits, genesis

geobotanical methods
Includes use on level 2 under mineral exploration(1).

geochemical cycle
Term introduced in 1982. Includes use on level 2 under geochemistry(1).

geochemical methods
Includes use on level 2 under mineral exploration(1).

geochemical prospecting *CR see under* geochemical methods *under* mineral exploration

geochemistry
Used for broad treatments as well as specific experiments. Includes use on level 1 (list A); on level 2 under major disciplines, area terms (list B), commodities (list C), intrusions(1), lava(1), organic materials(1), and rocks (list H, I, J). If 1, term set options are:
　processes
　　name of process [absorption, adsorption, chemical fractionation, chlorination, differentiation, diffusion, dissociation, electrolysis, endothermic reactions, fluorination, fractionation, hydration, hydrogenation, hydrolysis, ionization, ion exchange, nitrification, oxidation, photochemical reactions, photosynthesis, polymerization, pyrolysis, reduction, solution, substitution]
　properties
　　name of property [alkalinity, cation exchange capacity, Eh, electrochemical properties, pH, salinity, solubility, thermochemical properties, thermodynamic properties, etc.]
　topic [e.g. bibliography, classification, concepts, education, experimental studies, geochemical cycle, instruments, methods, practice, surveys, symposia, textbooks]
　subtopic

geochronology
For methods and topics other than those treated under absolute age, mainly for non-isotope methods. Includes use on level 1 (list A); on level 2 under age sets (list E), and area terms (list B). If 1, term set options are:
　methods
　　subtopic
　time scales
　age (list E)
　type of method [exposure age, fission-track dating, hydration of glass, lichenometry, optical mineralogy, paleomagnetism, particle-track dating (e.g. cosmic-ray track), racemization, radiation damage, tephrochronology, thermoluminescence, tree rings, varves]

subtopic
—— *CR see also* absolute age

geodes *CR see under* secondary structures *under* sedimentary structures

geodesy
Used infrequently and for geologic applications only. Includes use on level 1 (list A); on level 3 under Earth(1) gravity field(2). If 1, term set options are:
　topic [figure of Earth, geodetic coordinates, geoid, harmonics, methods, satellite measurements, surveys]
　subtopic

geodetic coordinates
Includes use on level 2 under geodesy(1).

geoid
Includes use on level 2 under geodesy(1). y
—— *CR see under* geodesy

geologic barometry
As of 1978, includes use on level 2 under fluid inclusions(1).

geologic hazards
A level 1 term as of 1978. Includes use on level 2 under impact statements(1) and nuclear facilities(1). Used for studies on the initiation, controls and effects of various geological phenomena which may be hazardous to human ecology. If 1, term set options are:
　avalanches
　　subtopic
　catastrophes
　　subtopic
　causes
　　subtopic
　controls
　　subtopic
　earthquakes
　　subtopic
　effects
　　subtopic
　explosions
　　subtopic
　human ecology
　　subtopic
　faults
　　subtopic
　floods
　　subtopic
　land subsidence
　　subtopic
　landslides
　　subtopic
　mudflows
　　subtopic
　observations
　　subtopic
　prediction
　　subtopic
　rock bursts

　subtopic
　site exploration
　　subtopic
　storms
　　subtopic
　sunspots
　　subtopic
　tsunamis
　　subtopic
　volcanoes
　　subtopic
—— *CR see also* earthquakes; land subsidence; *see also under* engineering geology *under* ____; *see also under* environmental geology *under* ____

geologic thermometry
As of 1978, includes use on level 2 under fluid inclusions(1).

geological methods
Includes use on level 2 under mineral exploration(1).

geology
For general treatments stressing the profession and the discipline. Includes use on level 1 (list A) and 2. If 1, term set options are:
　topic [education, history, nomenclature, philosophy, practice, principles, research, textbooks]
　subtopic
—— *see* areal geology; economic geology; engineering geology; environmental geology; extraterrestrial geology; glacial geology; marine geology; mathematical geology; mining geology; regional geology; structural geology

geometry
Includes use on level 2 under folds(1).
—— *see* plate geometry

geomorphologic effects
Includes use on level 2 under isostasy(1).

geomorphological methods
Includes use on level 2 under mineral exploration(1).

geomorphology
Includes use on level 1 (list A); on level 2 under area terms (list B), Moon (and other planets), and age terms (list E). If 1, term set options are:
　features [cryptoexplosion features, eolian features, erosion features, fluvial features, frost action, impact features, lacustrine features, mass movements, shore features, solution features, volcanic features]
　type of feature or topic [e.g. alluvial fans, beaches, caves, craters, deltas, drainage basins, drainage patterns, estuaries, evolution, karst, lagoons, meteor craters, patterned ground, quantitative geomorphology, reefs, rivers, terraces]

topic [applications, bibliography, concepts, education, environment, history, interpretation, landform description, landform evolution, maps, methods, practice, processes, symposia, textbooks]
subtopic
—— CR see also glacial geology
geophysical methods
Includes use on level 1 (list A); on level 2 under mineral exploration(1). Used for discussions which stress methodology of applied geophysics. If 1, term set options are:
type of method [acoustical methods, Earth-current methods, electrical methods, electromagnetic methods, gravity methods, infrared methods, magnetic methods, magnetotelluric methods, methods (for more than one method), radioactivity methods, remote sensing, seismic methods]
topic [applications, instruments, interpretation, techniques] or kind of platform [airborne methods, deep-tow methods, ground methods, marine methods, satellite methods], kind of method [induced polarization, resistivity, side-scanning methods, sonar methods]
—— CR see also geophysical surveys
geophysical surveys
For geophysical methods applied to specific areas. The methods are generally those of exploration and primarily concerned with the shallow structure of the Earth. Includes use on level 2 under area terms(1). See list B.
—— CR see acoustical surveys under geophysical surveys under ____; see Earth-current surveys under geophysical surveys under ____; see electrical surveys under geophysical surveys under ____; see electromagnetic surveys under geophysical surveys under ____; see gravity surveys under geophysical surveys under ____; see infrared surveys under geophysical surveys under ____; see magnetic surveys under geophysical surveys under ____; see magnetotelluric surveys under geophysical surveys under ____; see radioactivity surveys under geophysical surveys under ____; see seismic surveys under geophysical surveys under ____; see surveys under geophysical surveys under ____; see also geophysical methods
geophysics
For general treatments stressing the profession and the discipline plus

experimental studies on minerals and other materials. Includes use on level 1 (list A);on level 2 under bibliography(1), associations(1), education(1), symposia(1). If 1, term set options are:
topic [applications, bibliography, catalogs, classification, concepts, education, experimental studies, general, history, instruments, methods, nomenclature, objectives, observations, philosophy, practice, principles, symposia, textbooks, theoretical studies]
subtopic
—— CR see also deformation; engineering geology
—— see applied geophysics
Georgia
Term is used on level 1 as an area term (list O) only when referring to the United States. Included use for the Georgian Soviet Socialist Republic in Transcaucasia through 1977, but after 1977 use Georgian Republic for that.
geosynclines
Includes use on level 1 (list A). Used for large, mobile downwarping of the crust, which subsides as rocks and sediments accumulate to thicknesses of thousands of meters. Term set options are:
evolution
(name of geosyncline e.g. Alpine Geosyncline, Andean Geosyncline, Cordilleran Geosyncline, Hercynian Geosyncline, Mesogaea, Nevadan Geosyncline, Tethys) or subtopic
genesis
(name of geosyncline) or subtopic
processes
(name of geosyncline) or subtopic
—— CR see also orogeny; tectonics
geotechnics CR see engineering geology
geothermal energy
Includes use on level 1 as a commodity term (list C).
—— CR see also under economic geology under ____
geothermal gradient
Includes use on level 2 under heat flow(1).
—— CR see under heat flow
geothermics CR see heat flow
germanium
Includes use on level 1 and 2 as a chemical element (list D).
Germany
Country in Central Europe. Divided in 1949 into West Germany and East Germany. Includes use on level 1 as an area term (list O). See list B for term set options.

geysers
As of 1978, term is used on level 2 under thermal waters(1).
—— CR see under ____ under thermal waters
Ghana
Britain's former Gold Coast colony combined with British trust territory of Togo. Includes use on level 1 as an area term (list O). For term set options see list B.
Ginkgoales
Includes use on level 2 under gymnosperms(1). See list F.
glacial features
Includes use on level 2 under glacial geology(1).
glacial geology
Includes use on level 1 (list A) for geologic features and effects resulting from the action of glaciers and ice sheets. Term set options are:
ancient ice ages (older than Quaternary)
age
glacial features
name of feature [e.g. drumlins, eskers, fjords, fluting, glacial lakes, glacial polish, hanging valleys, kames, moraines, nunataks, sole marks, striations]
glaciation
topic [e.g. deglaciation, deposition, erosion, ice movement]
glaciers (for present-day glaciers)
subtopic
periglacial features
name of feature [e.g. extinct lakes, ice wedges, patterned ground, permafrost, pingos, solifluction]
topic [applications, bibliography, catalogs, classification, concepts, experimental studies, history, instruments, methods, nomenclature, philosophy, practice, symposia, textbooks, theoretical studies]
subtopic
—— CR see also geomorphology
glaciation
Includes use on level 2 under glacial geology(1); on level 3 under changes of level(1) and paleoclimatology(1).
—— CR see under glacial geology
glaciers
Includes use on level 2 under glacial geology(1) for present day glaciers; on level 2 under hydrology(1).
—— CR see under glacial geology
glasses
As of 1981, includes use on level 2 under igneous rocks(1). See list H. Before 1981, included use on level 3 under igneous rocks(1) pyroclastics and glasses(2).

—— *CR see under* igneous rocks
glauconite deposits
Term introduced in 1981. Includes use on level 1 and 2 as a commodity term (list C).

—— *CR see also under* economic geology *under* ——
global tectonics *CR see* plate tectonics
Globigerinacea
Includes use on level 2 under foraminifera(1). See list F.
glossaries
Includes use on level 1 (list A) for glossaries and atlases. Term set options are:
  topic [List B (except areal geology), or extraterrestrial geology, general, geophysics]
  subtopic

—— *CR see also* lexicons
Glossograptina
Includes use on level 2 under Graptolithina(1). See list F.
Glossopteridales
Term introduced in 1981. Includes use on level 2 under gymnosperms(1). See list F.
gneiss *CR see under* —— *under* metamorphic rocks
gneisses
Includes use on level 2 under metamorphic rocks(1). See list J.
Gnetales
Includes use on level 2 under gymnosperms(1). See list F.
gold
Includes use on level 1 and 2 as a chemical element (list D). As of 1981, use gold ores for gold as a commodity. Before 1981, included use on level 1 as a commodity term.
gold ores
Term introduced in 1981. Includes use on level 1 and 2 as a commodity term (list C).

—— *CR see also under* economic geology *under* ——
Gondwana
Theoretical ancient continent including India, Australia, Antarctica; and parts of southern Africa and South America. It is supposed to have fragmented and drifted apart in Post-Carboniferous time. Index countries and continents as applicable. Includes use on level 2 under continental drift (1).
Goniatitida
As of 1981, includes use on level 2 under Mollusca(1). See list F. Before 1981, included use on level 3 under Mollusca(1) Cephalopoda(2).
grabens *CR see under* —— *under* faults

grade
Includes use on level 2 under metamorphism(1).
graded bedding *CR see under* —— *under* sedimentary structures
gradient *see* geothermal gradient
granite deposits
Term introduced in 1981. Includes use on level 1 and 2 as a commodity term (list C).

—— *CR see also under* economic geology *under* ——
granites
Term introduced in 1981. Includes use on level 2 under igneous rocks(1). See list H.

—— *CR see under* igneous rocks
—— *see* alkali granites
granodiorite *CR see under* —— *under* igneous rocks
granodiorites
Term introduced in 1981. Includes use on level 2 under igneous rocks(1). See list H.

—— *CR see under* igneous rocks
granulites
Includes use on level 2 under metamorphic rocks(1). See list J.
graphite deposits
Term introduced in 1981. Includes use on level 1 and 2 as a commodity term (list C).

—— *CR see also under* economic geology *under* ——
graptolites
Term introduced in 1981 as the common name for Graptolithina. See list F. To be used for Graptolithina when they are used as tools in stratigraphy. Includes use on level 1. If 1, term set options are:
  biostratigraphy
  age [List E]
Graptolithina
Includes use on level 1 and 2 as a fossil term (list F).
Graptoloidea
Includes use on level 2 under Graptolithina(1). See list F.
gravel *CR see also under* —— *under* sediments
gravel deposits
Term introduced in 1981. Includes use on level 1 and 2 as a commodity term (list C). Includes sand when sand is used as a construction material.

—— *CR see also under* economic geology *under* ——
gravity field
Includes use on level 2 under Earth(1) and Moon(1).

—— *CR see under* Earth
gravity methods
Includes use on level 2 under geophysical methods(1).

gravity sliding
Includes use on level 2 under tectonics(1).
gravity surveys *CR see under* geophysical surveys *under* ——
Great Basin
Interior region between Sierra Nevada Mountains and S Cascade Range on W, and Wasatch Range and W face of Colorado Plateau on E. Introduced as a level 1 area term in 1978. Index states as applicable.
Great Britain
Index political divisions as applicable. Includes use on level 1 as an area term (list O). For term set options see list B.
Great Lakes
Largest group of fresh-water lakes in the world. Index lakes as applicable. Includes use on level 1 as an area term (list O). For term set options see list B.
Great Lakes region
Index Great Lakes, states, and provinces as applicable. Includes use on level 1 as an area term (list O). For term set options see list B.
Great Plains
A level 1 area term as of 1978. Sloping plateau extending from Rocky Mountains in W to the margin of the Central Plains in the U. S. and to the margin of the Laurentian Highlands in Canada. Index states and provinces as applicable. Until 1978, High Plains was used for the Great Plains from Nebraska southward.
Greater Antilles
Includes the major islands of the Antilles. As of 1977, includes use as a level 1 area term (list O). See list B for term set options. Index islands as applicable.
Greece
Includes use on level 1 as an area term (list O). For term set options see list B.
Greenland
Largest island in the world and a Danish province. Includes use on level 1 as an area term (list O). For term set options see list B.
greisen
Compositional term. As of 1978, term is used on level 2 under metasomatic rocks(1).
ground *see* patterned ground
ground ice
As of 1978, term is used on level 2 under permafrost(1).
ground water
For economically oriented papers, see water resources. This term used for subsurface waters, especially for studies on specific areas. Includes use on level 1 (list A) and 2. If 1, term set options are:

topic [age, aquifers, artesian waters, composition, connate waters, genesis, geochemistry, hydrodynamics, levels, models, movement, pollution, recharge, salt-water intrusion, surveys, systems analogs] (area)
subtopic (if no area is given)
—— *CR see also* hydrogeology; hydrology

group *see* amphibole group; chlorite group; epidote group; feldspar group; garnet group; humite group; melilite group; mica group; nepheline group; olivine group; pyroxene group; scapolite group; serpentine group; sodalite group; soil group; zeolite group

Guadeloupe
Combined islands of Basse-Terre and Grande-Terre which constitute an overseas department of France. Includes use on level 1 as an area term (list O). For term set options see list B.

Guatemala
Includes use on level 1 as an area term (list O). For term set options see list B.

guidebook *CR see under* —— *under* ——

Guinea
Formerly French Guinea. Includes use on level 1 as an area term (list O). For term set options see list B.

Guinea-Bissau
Formerly Portuguese Guinea. Includes use on level 1 as an area term (list O). For term set options see list B.

Gulf Coastal Plain
Index Mexican and U.S. states and Quintana Roo as applicable. Includes use on level 1 or 2 as an area term (list O). If 1, see term set options under list B.

Gulf of Aden
Between S coast of Arabian Peninsula and Somalia, E Africa. Includes use on level 1 or 2 as an area term (list O). If 1, see term set options under list B.

Gulf of California
Between peninsula of Baja California and the Mexican states of Sonora and Sinaloa. Includes use on level 1 as an area term (list O). For term set options see list B.

Gulf of Mexico
Relatively shallow oceanic-type basin encircled by Cuba, Mexico, and the United States. Includes use on level 1 or 2 as an area term (list O). If 1, see term set options under list B.

Guyana
Formerly British Guiana. Gained independence in 1966. Includes use on level 1 as an area term (list O). For term set options see list B.

Gymnocodiaceae
Term introduced in 1981. Includes use on level 2 under algae(1). See list F.

gymnosperm flora
Term introduced in 1982 as the common name for gymnosperms. See list F. To be used for gymnosperms when they are used as tools in stratigraphy. Includes use on level 1. If 1, term set options are:
biostratigraphy
age [list E]

gymnosperms
Includes use on level 1 and 2 as a fossil term (list F).

gypsum deposits
Term introduced in 1981. Includes use on level 1 and 2 as a commodity term (list C).
—— *CR see also under* economic geology *under* ——

habitat
Includes use on level 2 and 3 under fossil group(1). See list F.

hafnium
Includes use on level 1 and 2 as a chemical element (list D).

Haiti
Occupies western third of island of Hispaniola. Includes use on level 1 as an area term (list O). For term set options see list B.

Halecostomi
Term introduced in 1981. Includes use on level 2 under Pisces(1). See list F.

halides
Includes use on level 2 under minerals(1). See list L.
—— *CR see under* minerals

Hamamelididae
Term introduced in 1981. Includes use on level 2 under angiosperms(1). See list F.

harbors
As of 1978, term is used on level 2 under waterways(1).

harmonics
Includes use on level 2 under geodesy(1).

Hawaii
State including all of the Hawaiian Islands. Also southernmost and largest of the islands. Includes use on level 1 as an area term (list O). For term set options see list B. Also index Western U.S.

hazards *see* geologic hazards

heat flow
Includes use on level 1 (list A) and 2. Used for the product of the thermal

conductivity of a substance and the thermal gradient in the direction of the flow of heat. If 1, term set options are:
topic [anomalies, causes, changes, detection, distribution, experimental studies, genesis, geothermal gradient, heat sources, interpretation, measurement, observations, patterns, rates, regional patterns, temperature, theoretical studies, thermal conductivity]
subtopic (no area term)
—— *CR see also under* geophysical surveys *under* ——; *see also under* tectonophysics *under* ——

heat sources
Includes use on level 2 under heat flow(1).

heaving *see* frost heaving

heavy mineral deposits
Term introduced in 1981. Includes use on level 1 and 2 as a commodity term (list C).
—— *CR see also under* economic geology *under* ——

heavy minerals *CR see also* monazite; placers; titanium; zircon

Helicoplacoidea
Includes use on level 2 under Echinodermata(1). See list F.

helium
Includes use on level 1 and 2 as a chemical element (list D). As of 1981, use helium gas for helium as a commodity. Before 1981, included use on level 1 as a commodity term.

helium gas
Term introduced in 1981. Includes use on level 1 and 2 as a commodity term (list C).
—— *CR see also under* economic geology *under* ——

Hemichordata
Includes use on level 1 and 2 as a fossil term (list F). Term is to be used only when more specific terms do not apply.

Hemipteroida
Term introduced in 1981. Includes use on level 2 under Insecta(1). See list F.

Hepaticae
Includes use on level 2 under bryophytes(1). See list F.

Heterocorallia
Includes use on level 2 under Coelenterata(1). See list F.

Hexactiniaria
Includes use on level 2 under Coelenterata(1). See list F.

highways
Used for geological studies on highways and materials. A level 1 term as of 1978. Includes use on level 2 under foundations(1). If 1, term set options are:

construction
   subtopic
embankments
   subtopic
feasibility studies
   subtopic
foundations
   subtopic
frost action
   subtopic
materials, properties
   subtopic
planning
   subtopic
site exploration
   subtopic

Himalayas
Mountain system extending from Jammu and Kashmir in the W to Assam in the E. Index Tibet and countries as applicable. Includes use on level 1 as an area term (list O). For term set options see list B.

Hippomorpha
Term introduced in 1981. Includes use on level 2 under Mammalia(1). See list F.

history
Use term under various disciplines. Includes use as a level 2 or 3 term appropriate to a large number of topics, e.g. on level 2 under paleontology(1), geology(1), stratigraphy(1), and mineral exploration(1). See list G.

holmium
Includes use on level 1 and 2 as a chemical element (list D).

Holocene
Recent. Late Quaternary, from Pleistocene to present time. Includes use on level 1 as an age term (list E); on level 2 under paleo- terms, e.g. paleoecology, paleogeography, paleomagnetism; on level 3 under age terms(1). Above Pleistocene.

—— *CR see also under* geochronology *under* ——; *see also under* stratigraphy *under* ——

Holocephali
Term introduced in 1981. Includes use on level 2 under Pisces(1). See list F.

Holostei
As of 1981, includes use on level 2 under Pisces(1). See list F. Before 1981, included use on level 3 under Pisces(1) Osteichthyes(2).

Holothuroidea
Includes use on level 2 under Echinodermata(1). See list F.

Hominidae
As of 1981, includes use on level 2 under Mammalia(1). See list F. Before 1981, included use on level 3 under Mammalia(1) Primates(2).

Homo sapiens
As of 1981, includes use on level 2 under Mammalia(1). See list F. Before 1981, included use on level 3 under Mammalia(1) Primates(2).

Homoiostelea
Includes use on level 2 under Echinodermata(1). See list F.

Homostelea
Includes use on level 2 under Echinodermata(1). See list F.

Honduras
Includes use on level 1 as an area term (list O). For term set options see list B.

Hong Kong
British crown colony which includes the island of Hong Kong; and Kowloon Peninsula and New Territories on mainland. Includes use on level 1 as an area term (list O). For term set options see list B.

hornfels
Includes use on level 2 and 3 under metamorphic rocks(1). See list J.

hot springs
Includes use on level 2 under springs(1); on level 3 under thermal waters(1).

—— *CR see under* —— *under* springs; thermal waters

human ecology
As of 1978, term is used on level 2 under geologic hazards(1), land use(1), and pollution(1).

humates
Includes use on level 2 under organic materials(1).

humic acids
Includes use on level 2 under organic materials(1).

humite group
Includes use in combination with orthosilicates (i.e. orthosilicates, humite group) on level 2 under minerals(1). See list L.

Hungary
Includes use on level 1 as an area term (list O). For term set options see list B.

Hyalospongea
Includes use on level 2 under Porifera(1). See list F.

hydration of glass
Includes use on level 2 under geochronology(1).

—— *CR see under* geochronology

hydraulics
Used when discussing channels, or engineering and practical applications. For mathematical aspects of movement, use hydrodynamics. As of 1978, term is used on level 2 under shorelines(1), and waterways(1).

hydrocarbons
Includes use on level 2 under organic materials(1).

—— *CR see under* organic materials

hydrodynamics
Includes use on level 2 under ground water(1). Used when discussing mathematical aspects of ground water movement. For discussion of channels, engineering and practical applications, see hydraulics.

hydrogen
Includes use on level 1 and 2 as a chemical element (list D).

—— *CR see also* deuterium; tritium

hydrogeology
Used for general studies. For the science that deals with subsurface waters and related geologic aspects of surface waters. Includes use on level 1 (list A); on level 2 under area terms (list B). If 1, term set options are:
   topic [applications, bibliography, catalogs, classification, concepts, education, experimental studies, general, history, instruments, methods, nomenclature, objectives, philosophy, practice, symposia, textbooks, theoretical studies]
   subtopic

—— *CR see also* ground water; hydrology

hydrological methods
Includes use on level 2 under mineral exploration(1).

hydrology
Used for continental surface water. In cases where the interchange between ground and surface water is intimate, this set may be used with cycles on level 2. This covers cases of modern sediment transport. Includes use on level 1. Term set options are:
   cycles
      subtopic
   glaciers
      subtopic
   ice
      subtopic
   instruments
      name of instrument
         subtopic
   limnology
      subtopic
   methods
      subtopic
   atmospheric precipitation
      subtopic
   rivers and streams
      subtopic
   seepage

subtopic
snow
subtopic
surveys
(area)
techniques
name of technique or application
—— *CR see also* ground water; hydrogeology
hydrothermal alteration *CR see under* processes *under* metasomatism
Hydrozoa
Includes use on level 2 under Coelenterata(1). See list F.
Hymenopteroida
Term introduced in 1981. Includes use on level 2 under Insecta(1). See list F.
Hyolithes
Term introduced in 1981. Includes use on level 2 under Mollusca(1). See list F.
hypabyssal rocks
Term introduced in 1978. Includes use on level 2 under igneous rocks(1).
Hyracoidea
Term introduced in 1981. Includes use on level 2 under Mammalia(1). See list F.
Hystricomorpha
Term introduced in 1981. Includes use on level 2 under Mammalia(1). See list F.
ice
Includes use on level 2 under hydrology(1).
—— *CR see under* hydrology
—— *see* ancient ice ages; ground ice; sea ice
ice-rafting *CR see under* transport *under* sedimentation
Iceland
Independent country and island 155 miles SE of Greenland. Includes use on level 1 as an area term (list O). For term set options see list B. As of 1981, Atlantic Ocean is no longer autoposted.
ichnofossils
Includes use on level 1 and 2 as a fossil term (list F). As of 1981, is not to be used under sedimentary structures(1), and refers only to studies in which the emphasis is paleontology, not sedimentary petrology. For ichnofossils as components of sediments or sedimentary rocks, use lebensspuren.
Ichthyopterygia
Includes use on level 2 under Reptilia(1). See list F.
Ichthyosauria
Term introduced in 1981. Includes use on level 2 under Reptilia(1). See list F.

Idaho
Includes use on level 1 as an area term (list O). For term set options see list B.
ideal waves
Term introduced in 1978. Includes use on level 2 under ocean waves(1).
identification
Includes use as a level 2 or 3 term appropriate to a large number of topics, e.g. on level 2 under igneous rocks(1) and under sedimentary rocks(1). See list G.
igneous rocks
Used to describe crystalline rocks of primary origin (list H). Includes use on level 1 (list A); on level 2 under weathering(1), phase equilibria(1), and paragenesis(1). If 1, term set options are:
rock group [acidic composition, alkali basalts, alkali diorites, alkali gabbros, alkali granites, alkali syenites, alkalic composition, andesites, basalts, calc-alkalic composition, calcic composition, carbonatites, dacites, diabase, diorites, gabbros, glasses, granites, granodiorites, hypabyssal rocks, lamprophyres, mafic composition, monzonites, peridotites, phonolites, plutonic rocks, pyroclastics, quartz diorites, rhyodacites, rhyolites, syenites, trachyandesites, trachybasalts, trachytes, ultramafics, volcanic rocks]
rock name
topic [alteration, classification, composition, distribution, experimental studies, genesis, geochemistry, identification, melting, nomenclature, occurrence, petrography, petrology, properties, textures]
subtopic [e.g. chemical composition, mineral composition, physical properties]
—— *CR see also* magmas; metamorphic rocks; metasomatism; phase equilibria
ignimbrite *CR see under* —— *under* igneous rocks
Illinois
Includes use on level 1 as an area term (list O). For term set options see list B.
imagery
As of 1978, term is used on level 2 under remote sensing(1).
impact features
Includes use on level 2 under geomorphology(1).
—— *CR see under* geomorphology
impact statements
A level 1 term as of 1978. For official documents detailing the impact

of contemplated works or land use on the environment. Includes use on level 2 under land use(1), nuclear facilities(1), and pollution(1). If 1, term set options are:
conservation
subtopic
geologic hazards
subtopic
land use
subtopic
pollution
subtopic
reclamation
subtopic
waste disposal
subtopic
—— *CR see also under* environmental geology *under* ——
Inarticulata
Includes use on level 2 under Brachiopoda(1). See list F.
incertae sedis *CR see* problematic fossils
inclusions
Includes use on level 1 (list A) and 2 for studies on enclosures or enclaves in rocks or minerals. If 1, term set options are:
mineral inclusions
type of inclusion or host materials [e.g. name of mineral, name of rock, minerals, rocks] or topic
topic [analysis, changes, composition, detection, experimental studies, genesis, geologic barometry, geologic thermometry, P-T conditions, observations, theoretical studies]
subtopic
xenoliths
type of xenolith or host materials or topic
—— *CR see also* fluid inclusions
—— *see* fluid inclusions; mineral inclusions
India ·
Includes use on level 1 as an area term (list O). For term set options see list B.
Indian Ocean
Includes use on level 1 or 2 as an area term (list O). If 1, see term set options under list B.
—— *CR see also* Andaman Sea; Arabian Sea; Gulf of Aden; Gulf of Oman; Persian Gulf; Red Sea
Indian Ocean Islands
Term introduced in 1981. Includes use on level 1 as an area term (list O). For term set options see list B.
—— *CR see also* Comoro Islands; Malagasy Republic; Maldive Islands; Mauritius; Reunion; Seychelles
Indiana
Includes use on level 1 as an area term (list O). For term set options see list B.

indicators
Includes use on level 2 under paleo-climatology(1), under paleoecology(1), and under plate tectonics(1).

indium
Includes use on level 1 and 2 as a chemical element (list D).

Indochina
The SE peninsula of Asia including the Malay Peninsula. (French Indochina consisted of current Cambodia, Laos, and Vietnam). Index countries as applicable. Includes use on level 1 as an area term (list O). For term set options see list B.

Indonesia
Includes use on level 1 as an area term (list O). For term set options see list B.

industrial minerals
Includes use on level 1 as a commodity term (list C).
—— *CR see also* ceramic materials; corundum; feldspar; *see also under* economic geology *under* ——

industrial waste
As of 1978, term is used on level 2 under waste disposal(1).

inert gases *CR see* noble gases

infrared methods
Includes use on level 2 under geophysical methods(1). See spectroscopy(1) methods(2).

infrared surveys *CR see under* geophysical surveys *under* ——

Inocerami
Term introduced in 1981. Includes use on level 2 under Mollusca(1). See list F.

Insecta
Class. Includes use on level 1 and 2 as a fossil term (list F).

Insectivora
Includes use on level 2 under Mammalia(1). See list F.

instabilities *see* plasma instabilities

installations *see* marine installations; submarine installations; underground installations

instruments
Includes use as a level 2 or 3 term appropriate to a large number of topics, e.g. on level 2 under meteorology(1), geochemistry(1), heat flow(1), and hydrology(1). See list G.

intensity *see* seismic intensity

interactions *see* boundary interactions

interior
As of 1981, restricted to the Earth. Includes use on level 2 under Earth(1) and seismology(1). Before 1981, also included use on level 2 under Moon(1) and individual planets.

—— *see* lunar interior

interiors *see* planetary interiors

internal waves
Includes use on level 2 under ocean waves(1).

interplanetary dust
Includes use on level 2 under interplanetary space(1).

interplanetary space
Usually out-of-scope for GeoRef. Includes use on level 1. Term set options are:
comets
  subtopic
cosmic rays
  albedo
  cut-off rigidities
  intensity
  ionization
  subtopic (flares, measuring platform, particles, solar cycles)
  electromagnetic radiation
  subtopic
instruments
  name of instrument or platform or subtopic
interplanetary dust
  subtopic
shock waves
  subtopic
solar wind
  subtopic

interpretation
Includes use as a level 2 or 3 term appropriate to a large number of topics, e.g. on level 2 under foliation(1) and heat flow(1). See list G.

intrusion *see* salt-water intrusion

intrusions
Includes use on level 1 (list A). Used for the igneous rocks mass formed by emplacement of magma in pre-existing rock. It is also used for the process of emplacement. Term set options are:
kind of intrusion [batholiths, diatremes, dikes, laccoliths, layered intrusions, lopoliths, pipes, plugs, plutons, ring complexes, sills, stocks]
  topic
topic [age, classification, composition, contact metamorphism, distribution, emplacement, evolution, genesis, geochemistry, mechanism, occurrence, petrology]
  subtopic

—— *see* layered intrusions

inventory
Includes use on level 2 and 3 under commodity terms (list C); on level 3 under environmental geology and under engineering geology(1).

Invertebrata
Includes use on level 1 and 2 as a fossil term (list F); on level 3 under soils(1) biota(2). Term is to be used

on level 1 only when more specific terms do not apply or are too numerous to be recorded.
—— *CR see also* Annelida; Archaeocyatha; Arthropoda; Brachiopoda; Bryozoa; Coelenterata; Echinodermata; foraminifera; Graptolithina; Ichnofossils; Insecta; Mollusca; Ostracoda; Porifera; problematic fossils; Radiolaria; Trilobita; worms

invertebrates
Term introduced in 1981 as the common name for Invertebrata. See list F. To be used for Invertebrata when they are used as tools in stratigraphy. Includes use on level 1 when more specific common names do not apply or are too numerous to be recorded. If 1, term set options are:
biostratigraphy
  age [List E]

iodates
Includes use on level 2 under minerals(1). See list L.

—— *CR see under* minerals

iodides
Term introduced in 1981. Includes use on level 2 under minerals(1). See list L.

iodine
Includes use on level 1 and 2 as a chemical element (list D). As of 1981, use iodine deposits for iodine as a commodity. Before 1981, included use on level 1 as a commodity term.

iodine deposits
Term introduced in 1981. Includes use on level 1 and 2 as a commodity term (list C).

ion densities and temperatures
Includes use on level 2 under ionosphere(1).

ion exchange
As of 1978, term is used on level 2 under crystal chemistry(1).

Ionian Sea
Between SE coast of Italy and W Greece. Includes use on level 1 as an area term (list O). For term set options see list B.

ionization
Includes use on level 2 under aeronomy(1); on level 3 under interplanetary space(1) cosmic rays(2).

ionosphere
Usually out-of-scope for GeoRef. Includes use on level 1 for special bibliographies. Term set options are:
albedo
  [magnetic albedo, neutron albedo, also albedo of electromagnetic waves]
aurora
  subtopic

currents
  topic [e.g. field-aligned, ring, etc.]
disturbances
  subtopic
D-region
  subtopic
electrical field
  subtopic
electrojet
  equatorial regions
  polar regions
  subtopic
electron content
  topic [e.g. density, region of atmosphere, temperature]
E-region
  topic [e.g. density, sporadic E, temperature, variations]
F-region
  topic [density, spread-F, temperature]
instruments
  name of instrument or phenomena or platform
ion densities and temperatures
  name of ion or phenomenon or platform
  name of phenomena or platform
  name of region
  subtopic
particle precipitation
  name of particle or phenomenon or platform
  phenomenon or platform
scintillations
  subtopic
propagation
  name of phenomenon or subtopic
  type of wave or platform
ions
Includes use on level 2 under meteorology(1); on level 3 under geochemistry(1).
Iowa
Includes use on level 1 as an area term (list O). For term set options see list B.
Iran
Includes use on level 1 as an area term (list O). For term set options see list B.
Iraq
Includes use on level 1 as an area term (list O). For term set options see list B.
Ireland
The Republic of Ireland, also called Eire, which occupies the 26 counties in the S, central, and NW of the island of Ireland. Includes use on level 1 as an area term (list O). For term set options see list B.
iridium
Includes use on level 1 and 2 as a chemical element (list D).

Irish Sea
Between England and Ireland. Includes use on level 1 as an area term (list O). For term set options see list B.
iron
Includes use on level 1 and 2 as a chemical element (list D). As of 1981, use iron ores for iron as a commodity. Before 1981, included use on level 1 as a commodity term.
iron ores
Term introduced in 1981. Includes use on level 1 and 2 as a commodity term (list C).
—— CR see also under economic geology under ——
irregularities see bedding plane irregularities
irrigation
As of 1978, term is used on level 2 under waterways(1).
island arcs
Includes use on level 2 under plate tectonics(1).
islands see barrier islands
isostasy
To be used for the condition of equilibrium of the crust above the mantle, and related features. Includes use on level 1 (list A). Term set options are:
  topic [anomalies, causes, changes, compensation, concepts, detection, genesis, geomorphologic effects, interpretation, mechanism, observations]
  subtopic (no area term)
—— CR see also changes of level; neotectonics; see also under structural geology under ——; see also under tectonophysics under ——
isotope dating CR see absolute age
isotopes
Includes use on level 1 (list A); on level 2 under chemical elements (list D). Used for the application of the study of radioactive and stable isotopes, especially their abundances, to geology, excluding absolute age dating. If 1, term set options are:
  element
  specific isotope or ratio
  material (first-level terms only)
  type of material
  topic [abundance, analysis, fractionation, methods, ratios, tracers]
  subtopic
—— CR see also absolute age; geochronology
Israel
Includes use on level 1 as an area term (list O). For term set options see list B.

Italy
Includes use on level 1 as an area term (list O). For term set options see list B.
Ivory Coast
Includes use on level 1 as an area term (list O). For term set options see list B.
Jamaica
Includes use on level 1 as an area term (list O). For term set options see list B.
Jan Mayen
Norwegian volcanic island between Greenland and Norway. Includes use on level 1 as an area term (list O). For term set options see list B.
Japan
Includes use on level 1 as an area term (list O). For term set options see list B.
Japan Sea
Between Japan in the E; and Korea and Primorye Kray in the Soviet Far East on the W. Includes use on level 1 as an area term (list O). For term set options see list B.
joints CR see under —— under fractures
Jordan
Includes use on level 1 as an area term (list O). For term set options see list B.
Jupiter
Includes use on level 1 and 2. See term set options under Moon(1).
Jurassic
Includes use on level 1 as an age term (list E); on level 2 under paleoterms, e.g. paleoecology, paleogeography, paleomagnetism. Above Triassic, below Cretaceous.
—— CR see also under geochronology under ——; see also under stratigraphy under ——
kames CR see under glacial features under glacial geology
Kansas
Includes use on level 1 as an area term (list O). For term set options see list B.
kaolin deposits
Term introduced in 1981. Includes use on level 1 and 2 as a commodity term (list C).
—— CR see also under economic geology under ——
karst CR see under solution features under geomorphology
Kentucky
Includes use on level 1 as an area term (list O). For term set options see list B.
Kenya
Includes use on level 1 as an area term (list O). For term set options see list B.

kerogen
Includes use on level 2 under organic materials(1).

Kirkbyocopina
Term introduced in 1981. Includes use on level 2 under Ostracoda(1). See list F.

Korea
A peninsula on E coast of Asia, since 1948 partitioned into two republics. Includes use on level 1 as an area term (list O). For term set options see list B. Use North Korea and/or South Korea on level 3.

krypton
Includes use on level 1 and 2 as a chemical element (list D).

Kuwait
Includes use on level 1 as an area term (list O). For term set options see list B.

Labrador
Includes use on level 1 as an area term (list O). For term set options see list B.

Labyrinthodontia
Includes use on level 2 under Amphibia(1). See list F.

laccoliths
Includes use on level 2 under intrusions(1).

—— *CR see under* intrusions

lacustrine features
Includes use on level 2 under geomorphology(1).

Lagomorpha
Includes use on level 2 under Mammalia(1). See list F.

lagoons *CR see under* —— *under* ecology; *see under* environment *under* sedimentation

lakes *CR see under* —— *under* ecology; *see under* environment *under* sedimentation; *see under* lacustrine features *under* geomorphology; *see under* limnology *under* hydrology

lamprophyres
Term introduced in 1981. Includes use on level 2 under igneous rocks(1). See list H.

—— *CR see under* igneous rocks

land subsidence
A level 1 term as of 1978. Includes use on level 2 under foundations(1) and geologic hazards(1). Used for geotechnical studies on land subsidence. If 1, term set options are:
causes
　subtopic
controls
　subtopic
foundations
　subtopic
mines
　subtopic

site exploration
　subtopic
settlement
　subtopic
stability
　subtopic
solution features
　subtopic

land use
A level 1 term as of 1978. Includes use on level 2 under impact statements(1). If 1, term set options are:
agriculture
　subtopic
arid environment
　subtopic
changes
　subtopic
classification
　subtopic
conservation
　subtopic
controls
　subtopic
effects
　subtopic
human ecology
　subtopic
impact statements
　subtopic
inventory
　subtopic
legislation
　subtopic
management
　subtopic
maps
　cartography
mines
　subtopic
natural resources
　subtopic
planning
　subtopic
preservation
　subtopic
reclamation
　subtopic
recreation
　subtopic
regional planning
　subtopic
remote sensing
　subtopic
soils
　subtopic
underground space
　subtopic
urban planning
　subtopic

—— *CR see also* conservation; pollution; reclamation; *see also under* environmental geology *under* ——

landform description
Includes use on level 2 under geomorphology(1).

—— *CR see under* geomorphology

landform evolution
Includes use on level 2 under geomorphology(1).

landslides
As of 1978, term is used on level 2 under geologic hazards(1), and slope stability(1).

—— *CR see under* —— *under* geologic hazards; geomorphology; *see under* slope stability

lanthanum
Includes use on level 1 and 2 as a chemical element (list D).

Laos
Includes use on level 1 as an area term (list O). For term set options see list B.

laterites *CR see under* —— *under* soils

Laurasia
The protocontinent of the Northern Hemisphere which is a combination of Laurentia and Eurasia. It included most of North America, Greenland, and much of Eurasia excluding India. Includes use on level 2 under continental drift(1).

lava
This set is used for all molten extrusives and the rocks that solidify from them. Includes use on level 1 (list A); on level 2 under isotopes(1). If 1, term set options are:
topic [age, alteration, analysis, classification, composition, distribution, flow mechanism, genesis, geochemistry, nomenclature, observations, occurrence, petrology, properties, temperature, viscosity]
　subtopic [e.g. electrical properties, magnetic properties, pillow lava, pillow structure]

—— *CR see also* igneous rocks; magmas

layered intrusions
Not a valid term from 1975 through 1977. After 1977, includes use on level 2.

lead
Includes use on level 1 and 2 as a chemical element (list D). As of 1981, use lead ores for lead as a commodity. Before 1981, included use on level 1 as a commodity term.

lead ores
Term introduced in 1981. Includes use on level 1 and 2 as a commodity term (list C).

—— *CR see also under* economic geology *under* ——

lead-zinc deposits
As of 1978, term includes use on level 1 and 2 as a commodity (list C).

Lebanon
Includes use on level 1 as an area term (list O). For term set options see list B.

legislation
As of 1978, term is used on level 2 under land use(1).

Leperditicopida
Includes use on level 2 under Ostracoda(1). See list F.

Lepidocystoidea
Includes use on level 2 under Echinodermata(1). See list F.

Lepidopteroida
Term introduced in 1981. Includes use on level 2 under Insecta(1). See list F.

Lepidosauria
Includes use on level 2 under Reptilia(1). See list F.

Lepospondyli
Includes use on level 2 under Amphibia(1). See list F.

Lesotho
Former British colony of Basutoland. Includes use on level 1 as an area term (list O). For term set options see list B.

Lesser Antilles
The smaller islands of the Antilles extending in an arc from Puerto Rico to the islands N of Venezuela. Index Barbados, Trinidad and island groups as applicable. As of 1977, includes use as a level 1 area term (list O). See list B for term set options.

level *see* changes of level

levels
Includes use on level 2 under ground water(1).

lexicons
Includes use on level 1 (list A) for systematic arrangement of words in a particular language, or of a considerable number of them, and their definition. Term set options are:
topic (List B or extraterrestrial geology or geophysics)
language (if not English)
—— *CR see also* glossaries

Liberia
Includes use on level 1 as an area term (list O). For term set options see list B.

Libya
Includes use on level 1 as an area term (list O). For term set options see list B.

lichenometry
Includes use on level 2 under geochronology(1).
—— *CR see under* geochronology

lichens
Includes use on level 1 and 2 as a fossil term (list F).

Lichida
Includes use on level 2 under Trilobita(1). See list F.

life origin
Term introduced in 1978. Includes use on level 2 under paleontology(1).

lignite
Includes use on level 1 and 2 as a commodity term (list C); on level 3 under sedimentary rocks(1) organic residues(2). See list I.
—— *CR see also under* —— under sediments; *see also under* economic geology *under* ——

Liliidae
Term introduced in 1981. Includes use on level 2 under angiosperms(1). See list F.

limestone *CR see also under* —— *under* sedimentary rocks

limestone deposits
Term introduced in 1981. Includes use on level 1 and 2 as a commodity term (list C).
—— *CR see also under* economic geology *under* ——

limnology
Includes use on level 2 under hydrology(1); on level 3 under geomorphology(1) lacustrine features(2).
—— *CR see under* hydrology

lineation
Small-scale features. Includes use on level 1 (list A); on level 2 and 3 under structural analysis(1). If level 1, term set options are:
experimental studies
subtopic
genesis
subtopic
interpretation
subtopic
style
subtopic [e.g. boudinage, elongate minerals, flow lines, mullions, slickensides]
—— *CR see also* foliation; structural analysis

liquefaction
Not to be used under phase equilibria. As of 1981, includes use on level 2 under slope stability(1).

liquid waste
As of 1978, term is used on level 2 under waste disposal(1).

Lissamphibia
Includes use on level 2 under Amphibia(1). See list F.

lithium
Includes use on level 1 and 2 as a chemical element (list D). As of 1981, use lithium ores for lithium as a commodity. Before 1981, included use on level 1 as a commodity term.

lithium ores
Term introduced in 1981. Includes use on level 1 and 2 as a commodity term (list C).

—— *CR see also under* economic geology *under* ——

lithofacies
Includes use on level 2 under sedimentary rocks(1), sediments(1), and reefs(1).

lithostratigraphy
Includes use on level 2 under sedimentary rocks(1), metamorphic rocks(1) and sediments(1).

Litopterna
Includes use on level 2 under Mammalia(1). See list F.

Lituolacea
Includes use on level 2 under foraminifera(1). See list F.

logging *see* acoustical logging; caliper logging; dipmeter logging; electrical logging; electromagnetic logging

lopoliths
Includes use on level 2 under intrusions(1).
—— *CR see under* intrusions

Louisiana
Includes use on level 1 as an area term (list O). For term set options see list B.

low-velocity zones
Term introduced in 1978 as a general term, i.e. applicable to crust, mantle, etc. As of 1981, includes use on level 2 under mantle(1).

lunar crust
Term introduced in 1981. Includes use on level 2 under Moon(1).

lunar interior
Term introduced in 1981. Includes use on level 2 under Moon(1).

lunar studies *CR see* Moon

lutetium
Includes use on level 1 and 2 as a chemical element (list D).

Luxembourg
Includes use on level 1 as an area term (list O). For term set options see list B.

Lycopsida
Includes use on level 2 under pteridophytes(1). See list F.

Lytoceratida
Term introduced in 1981. Includes use on level 2 under Mollusca(1). See list F.

Machaeridia
Includes use on level 2 under Echinodermata(1). See list F.

Macroscelida
Term introduced in 1981. Includes use on level 2 under Mammalia(1). See list F.

Madagascar *CR see* Malagasy Republic

Madeira
Island group belonging to Portugal 600 miles SW of Lisbon and 400 miles W of Morocco. Includes use on level 1 as an area term (list O). For term set options see list B.

mafic composition
Term introduced in 1978. Includes use on level 2 under igneous rocks(1).

magmas
Includes use on level 1 (list A); on level 2 under phase equilibria(1). Used for naturally occurring mobile rock material, generated within the Earth and capable of intrusion and extrusion. If 1, term set options are: topic [age, classification, composition, contamination, differentiation, evolution, genesis, geochemistry, properties, temperature, viscosity]
subtopic
—— CR see also igneous rocks; intrusions; lava

magnesite deposits
Term introduced in 1981. Includes use on level 1 and 2 as a commodity term (list C).
—— CR see also under economic geology under ——

magnesium
Includes use on level 1 and 2 as a chemical element (list D). As of 1981, use magnesium ores for magnesium as a commodity. Before 1981, included use on level 1 as a commodity term.

magnesium ores
Term introduced in 1981. Includes use on level 1 and 2 as a commodity term (list C).

magnetic domains
Level 2 term introduced in 1978 under paleomagnetism(1).

magnetic field
Includes use on level 2 under astrophysics and solar physics(1), aurora(1), Earth(1), Moon(1), and any of the planets.
—— CR see under Earth

magnetic methods
Includes use on level 2 under geophysical methods(1).

magnetic storms
Includes use on level 2 under magnetosphere(1); on level 3 under aurora(1) magnetic field(2).

magnetic surveys CR see under geophysical surveys under ——

magnetic tail
Includes use on level 2 under magnetosphere(1).

magnetopause
Includes use on level 2 under magnetosphere(1).

magnetosheath
Includes use on level 2 under magnetosphere(1).

magnetosphere
Includes use on level 1 for special indexes. Term set options are:

bow shock waves
topic
coordinate systems
topic
cosmic rays
albedo
cutoff rigidities
intensity
electrical field
topic or platform
general
subtopic
instruments
name of instrument
magnetic tail
topic or platform
magnetic storms
topic [e.g. initial phase, main phase, plasmasphere, recovery phase, substorms, sudden-commencements, trapped particles]
magnetopause
topic or platform
magnetosheath
topic or platform
plasma instabilities
topic or platform
plasma motion
topic [circulation, convection]
plasmapause
topic or platform
solar wind
subtopic
techniques
name of technique
type of phenomena
trapped particles
type of particle
type of phenomena
variations
topic [e.g. micropulsations, solar eclipses]
propagation
subtopic
type of wave
whistlers
(platform if satellite)
topic

magnetotelluric methods
Includes use on level 2 under geophysical methods(1).

magnetotelluric surveys CR see under geophysical surveys under ——

magnitude
Includes use on level 2 under earthquakes(1). As of 1981, restricted to magnitude of earthquakes.

Magnoliidae
Term introduced in 1981. Includes use on level 2 under angiosperms(1). See list F.

Maine
Includes use on level 1 as an area term (list O). For term set options see list B.

maintenance
As of 1978, term is used on level 2 under reservoirs(1).

Malacostraca
Includes use on level 2 under Arthropoda(1). See list F. As of 1981, used for Decapoda.

Malagasy Republic
Formerly Madagascar. Became Malagasy Republic in 1958. Includes use on level 1 as an area term (list O). For term set options see list B.

Malawi
Formerly Nyasaland. At one time it was part of British Central African Protectorate and later part of the Federation of Rhodesia and Nyasaland. Includes use on level 1 as an area term (list O). For term set options see list B.

Malay Archipelago
Largest island group in world off SE coast of Asia between Pacific and Indian oceans. Index countries as applicable. Includes use on level 1 or 2 as an area term (list O). If 1, see term set options under list B.
—— CR see also New Guinea

Malaysia
Or officially known as Federation of Malaysia. Independent federation, SE Asia, consisting of eleven states (West Malaysia) on the Malay Peninsula and two states (East Malaysia) on the island of Borneo. Includes use on level 1 as an area term (list O). For term set options see list B.

Maldive Islands
Group of 19 atolls in the Indian Ocean 300 miles SW of southern tip of India. Includes use on level 1 as an area term (list O). For term set options see list B.

Mali
Includes use on level 1 as an area term (list O). For term set options see list B.

Malta
Comprises 3 islands. Includes use on level 1 as an area term (list O). For term set options see list B.

Mammalia
Includes use on level 1 and 2 as a fossil term (list F).

mammals
Term introduced in 1981 as the common name for Mammalia. See list F. To be used for Mammalia when they are used as tools in stratigraphy. Includes use on level 1. If 1, term set options are:
biostratigraphy
age [List E]

man see fossil man

man, fossil CR see fossil man

management
Used as a general term. As of 1978, term is used on level 2 under land use(1) and shorelines(1).

Mandibulata
Term introduced in 1981. Includes use on level 2 under Arthropoda(1). See list F.

manganese
Includes use on level 1 and 2 as a chemical element (list D). As of 1981, use manganese ores for manganese as a commodity. Before 1981, included use on level 1 as a commodity term.

manganese ores
Term introduced in 1981. Includes use on level 1 and 2 as a commodity term (list C).

Manitoba
Includes use on level 1 as an area term (list O). For term set options see list B.

mantle
Includes use on level 1 (list A); on level 2 under seismology(1). Used for the zone of Earth below the crust and above the core (to a depth of 3480 km). If 1, term set options are: topic [age, composition, earthquakes, elastic waves, evolution, genesis, geochemistry, interpretation, low-velocity zones, processes, properties, temperature]
subtopic (no area term)
——— CR see also under seismology under ———; see also under tectonophysics under ———

maps
Includes use on level 1 (list A); on level 2 under geomorphology(1), land use(1), and environmental geology(1); on level 3 under area(1) discipline from list B(2). If 1, term set options are:
cartography [for method, instrument, program]
subtopic
topic (for global maps; List B except for areal geology)
subtopic (no area term) e.g. type of map [e.g. economic geology, environmental geology, geologic maps, geomorphologic maps, hydrogeologic maps, photogeologic maps, soils maps, surficial geology, tectonic maps]
——— CR see also under areal geology under ———; see also under economic geology under ———; see also under engineering geology under ———; see also under environmental geology under ———; see also under general under ———; see also under geochemistry under ———; see also under geochronology under ———; see also under geomor-

phology under ———; see also under geophysical surveys under ———; see also under hydrogeology under ———; see also under oceanography under ———; see also under seismology under ———; see also under soils under ———; see also under stratigraphy under ———; see also under structural geology under ———; see also under tectonophysics under ———

marble CR see also under ——— under metamorphic rocks

marble deposits
Term introduced in 1981. Includes use on level 1 and 2 as a commodity term (list C).
——— CR see also under economic geology under ———

marbles
Includes use on level 2 under metamorphic rocks(1). See list J. As of 1981, use marble deposits for marble as a commodity.

Mariana Islands
Group of islands 1500 miles E of Philippine Islands. The islands, not including Guam, formerly were part of the U.S. Trust Territory of the Pacific Islands. Commonwealth status within the U.S. was achieved on March 24, 1976. Includes use on level 1 as an area term (list O). For term set options see list B.

marine geology
Includes use on level 1 (list A). Used for that aspect of the study of the ocean which deals specifically with the ocean floor and the ocean-continent border. Term set options are:
topic [applications, bibliography, bottom features, catalogs, concepts, education, experimental studies, geochemistry, history, instruments, methods, nomenclature, objectives, observations, practice, principles, textbooks, theoretical studies]
subtopic (no area term)
——— CR see also continental shelf; continental slope; oceanography; see also under oceanography under ———

marine installations
Term introduced in 1976 and used on level 1 beginning in 1978. Before 1978, included use on level 2 under engineering geology(1). Term set options are:
construction
subtopic
design
subtopic
experimental studies
subtopic
feasibility studies

subtopic
foundations
subtopic
gravity platforms
subtopic
instruments
subtopic
marine platforms
subtopic
methods
subtopic
pipelines
subtopic
site exploration
subtopic
submarine installations
subtopic
theoretical studies
subtopic
——— CR see also under engineering geology under ———

marine sediments
Term introduced in 1981. Includes use on level 2 under sediments(1). See list N.

Maritime Provinces
Often called the Maritimes. Index provinces as applicable. Introduced as level 1 area term in 1978.

Mars
Includes use on level 1 and 2. See entry under Moon(1) for term set options.

Marshall Islands
Group of 34 atolls and coral islands, including Eniwetok and Kwajalein, SW of Hawaii and E of Guam. Part of the U.S. Trust Territory of the Pacific Islands. Includes use on level 1 as an area term (list O). For term set options see list B.

Marsupialia
Includes use on level 2 under Mammalia(1). See list F.

Maryland
Includes use on level 1 as an area term (list O). For term set options see list B.

mass movements
Includes use on level 2 under geomorphology(1).
——— CR see under geomorphology; slope stability

Massachusetts
Includes use on level 1 as an area term (list O). For term set options see list B.

Mastodontoidea
Term introduced in 1981. Includes use on level 2 under Mammalia(1). See list F.

materials
Includes use on level 2 under diagenesis(1) and in combination with properties (i.e. materials, properties) on level 2 under engineering geology(1), rock mechanics(1), and soil mechanics(1).

—— *see* ceramic materials; construction materials; organic materials; standard materials

mathematical geology
For mathematics as applied to geology. Includes use on level 1 (list A). Term set options are:
topic [bibliography, concepts, education, history, interpretation, methods, nomenclature, objectives, philosophy, principles, symposia, textbooks, theoretical studies]
  subtopic
—— *CR see also* automatic data processing

Mauritania
Includes use on level 1 as an area term (list O). For term set options see list B.

Mauritius
An independent state in the Mascarene Islands about 450 miles E of the Malagasy Republic. Includes use on level 1 as an area term (list O). For term set options see list B.

measurement
Includes use on level 2 under heat flow(1).

mechanics
Includes use on level 2 under faults(1) and folds(1); on level 3 under plate tectonics(1).
—— *see* rock mechanics; soil mechanics

mechanism
Includes use as a level 2 or 3 term appropriate to a large number of topics, e.g. on level 2 under orogeny(1), under salt tectonics(1), and under plate tectonics(1). See list G.
—— *see* flow mechanism; focal mechanism

Mecopteroida
Term introduced in 1981. Includes use on level 2 under Insecta(1). See list F.

Mediterranean region
The Mediterranean Sea and its islands, the Adriatic Sea, and parts of those countries along their shores in S Europe, N Africa, and the Middle East. Index Adriatic Sea, Mediterranean Sea, countries, and islands as applicable. Includes use on level 1 or 2 as an area term (list O). If 1, see term set options under list B.
—— *CR see also* Corsica; Sardinia

Mediterranean Sea
Enclosed by Europe on the W and N, Asia on the E, and Africa on the S. Includes use on level 1 or 2 as an area term (list O). If 1, see term set options under list B. As of 1981, includes the Black Sea.
—— *CR see also* Adriatic Sea; Aegean Sea; Balearic Islands; Corsica; Ionian Sea; Malta; Sardinia; Tyrrhenian Sea .

meetings *CR see* symposia

megaspores
Includes use on level 2 under palynomorphs(1). See list F.

Melanesia
Collective name for islands NE of Australia. Index island groups as applicable. Includes use on level 1 and 2 as an area term (list O). For term set options see list B.
—— *CR see also* Fiji; New Hebrides; Solomon Islands

melange *CR see under* —— *under* ——

melilite group
Includes use in combination with orthosilicates (i.e. orthosilicates, melilite group) on level 2 under minerals(1). See list L.

melting
Includes use on level 2 under phase equilibria(1).

mercury
Includes use on level 1 and 2 as a chemical element (list D). As of 1981, use mercury ores for mercury as a commodity. Use Mercury Planet for the planet. Before 1981, included use on level 1 as a commodity term.

mercury ores
Term introduced in 1981. Includes use on level 1 and 2 as a commodity term (list C).
—— *CR see also under* economic geology *under* ——

Mercury Planet
Includes use on level 1 and 2. See entry under Moon(1) for term set options.

Merostomata
Includes use on level 2 under Arthropoda(1). See list F.

Mesogastropoda
Term introduced in 1981. Includes use on level 2 under Mollusca(1). See list F.

Mesosauria
Term introduced in 1981. Includes use on level 2 under Reptilia(1). See list F.

Mesozoic
Includes use on level 1 as an age term (list E); on level 2 under paleoterms, e.g. paleoecology, paleogeography, paleomagnetism. From the end of the Paleozoic to the beginning of the Cenozoic.
—— *CR see also* Cretaceous; Jurassic; Triassic; *see also under* geochronology *under* ——; *see also under* stratigraphy *under* ——

metaigneous rocks
Includes use on level 2 under metamorphic rocks(1). See list J.

metal ores
Term introduced in 1981. Includes use on level 1 and 2 as a commodity term (list C).
—— *CR see also* aluminum ores; antimony ores; arsenic ores; bauxite; beryllium ores; bismuth ores; chromite ores; cobalt ores; gold ores; iron ores; lead ores; lithium ores; magnesium ores; manganese ores; mercury ores; molybdenum ores; nickel ores; niobium ores; platinum ores; pyrite ores; silver ores; tantalum ores; thorium ores; tin ores; titanium ores; tungsten ores; uranium ores; zinc ores; *see also under* economic geology *under* ——

metals
Includes use on level 1 and 2 as a chemical element (list D). As of 1981, use metal ores for metals as a commodity. Before 1981, included use on level 1 as a commodity term.

metamorphic rocks
Includes use on level 1 (list A); on level 2 under paragenesis(1), phase equilibria(1), and weathering(1). If 1, term set options are:
rock group [amphibolites, cataclasites, eclogite, garnetite, gneisses, granulites, hornfels, itabirite, marbles, metaigneous rocks, metaplutonic rocks, metasedimentary rocks, metavolcanic rocks, migmatites, mylonites, phyllites, phyllonites, quartzites, schists, slates]
  rock name
topic [Used for general studies where several rock types are discussed or where rock is not assignable to one of the groups above; age, classification, composition, correlation, distribution, evolution, experimental studies, facies, genesis, geochemistry, lithostratigraphy, mineral assemblages, occurrence, petrography, petrology, properties, textures]
  subtopic
—— *CR see also* igneous rocks; metamorphism; metasomatism

metamorphism
Includes use on level 1 (list A). Term set options are:
kind of metamorphism [anchimetamorphism, autometamorphism, burial metamorphism, contact metamorphism, dynamic metamorphism, grade, polymetamorphism, prograde metamorphism, regional metamorphism, retrograde metamorphism, shock metamorphism, thermal metamorphism, ultrametamorphism]
  topic (high-grade metamorphism, low-grade metamorphism)

topic [For more than one kind or for a kind not assignable to one of the categories above: age, causes, classification, concepts, distribution, environment, evolution, genesis, mechanism, migration of elements, processes, P-T conditions, rates, rheomorphism, temperature, theoretical studies, zoning]
   subtopic
—— see burial metamorphism; contact metamorphism; dynamic metamorphism; prograde metamorphism; regional metamorphism; retrograde metamorphism; shock metamorphism; thermal metamorphism

metaplutonic rocks
Term introduced in 1978. Includes use on level 2 under metamorphic rocks(1). See list J.

metasedimentary rocks
Includes use on level 2 under metamorphic rocks(1). See list J.

metasomatic rocks
As of 1978, term is used on level 1 for any rock produced by replacement processes at constant volume with little disturbance of textural or structural features. Includes use on level 3 under petrology(1) and under metasomatism(1). Term set options are:
   rock group [greisen, propylite, skarn]
   rock name or process or subtopic
   topic [age, classification, composition, distribution, evolution, facies, genesis, geochemistry, mineral assemblages, occurrence, petrography, petrology, textures]
   subtopic

metasomatism
Includes use on level 1 (list A). Term set options are:
   materials
   name of material [e.g. carbonate rocks, granite, igneous rocks, sedimentary rocks]
   processes
   name of process [e.g. albitization, alunitization, analcimization, argillization, dolomitization, granitization, greisenization, hydrothermal alteration, kaolinization, laumonitization, microclinization, muscovitization, palagonitization, propylitization, pyrometasomatism, scapolitization, serpentinization, silicification, zeolitization]
topic [age, causes, environment, experimental studies, geochemistry, interpretation, mechanism, rates, theoretical studies]
   subtopic

metavolcanic rocks
Term introduced in 1978. Includes use on level 2 under metamorphic rocks(1). See list J.

meteor craters
Includes use on level 1 (list A); on level 3 under area(1) geomorphology(2) and under geomorphology(1) impact features(2). Used for impact craters formed by the falling of a large meteorite onto a surface. If 1, term set options are:
   topic [age, causes, classification, concepts, distribution, evolution, genesis, interpretation, observations, occurrence, patterns]
   subtopic
—— *CR see also* meteorites; *see also under* geomorphology *under*

meteorite flux
Term introduced in 1981. Includes use on level 2 under meteorites(1).

meteorites
Includes use on level 1 (list A) and 2. Used for any meteoroid that has fallen to the Earth's surface in one piece or in fragments without being completely vaporized. Term set options are:
   topic [age, bibliography, catalogs, collections, composition, detection, distribution, experimental studies, meteorite flux, genesis, geochemistry, interpretation, isotopes, organic materials, phase equilibria, properties, radioactivity, textures]
   name of meteorite or type [e.g. achondrites, chondrites, howardite, iron meteorites, stony irons] or subtopic
—— *CR see also* meteor craters; tektites

meteorology
Primarily used for special indexes. Includes use on level 1. Term set options are:
   aerosols
   topic [composition, convection, particles, radioactive tracers, turbidity]
   boundary interactions
   topic
   circulation
   topic [meridional circulation, zonal circulation, etc.]
   composition
   elements or compound or region of atmosphere
   convection
   topic
   diffusion
   models
   property diffused
   type of diffusion
   electromagnetic waves

   topic [currents, lightning, raindrops, space charge, thunderstorms]
   experimental studies
   topic
   instruments
   name of instrument
   type of phenomenon
   ions
   name of ion
   particles
   name of particle or topic or subtopic
   models
   storms
   models
   type of storm
   techniques
   type of phenomenon
   technique
   temperature
   topic
   theoretical studies
   topic
   turbulence
   area
   subtopic
   topic
   water
   topic [atmospheric precipitation, clouds, droplets, humidity, isotopes, raindrops, name of isotope]
   waves
   type of wave
   winds
   type of wind

methods
This term applies to theoretical and experimental studies, while the term techniques is used for discussion of samples. Includes use as a level 2 or 3 term appropriate to a large number of topics, e.g. on level 2 under absolute age(1), geochemistry(1), geophysical methods(1) organic materials(1), and soils(1) and mineral exploration(1). See list G.
—— see acoustical methods; biogeochemical methods; electrical methods; electromagnetic methods; geobotanical methods; geochemical methods; geological methods; geomorphological methods; geophysical methods; gravity methods; hydrological methods; infrared methods; magnetic methods; magnetotelluric methods; photogeologic methods; radioactivity methods; seismic methods

Mexico
Includes use on level 1 or 2 as an area term (list O). If 1, see term set options under list B.
—— *CR see also* Gulf Coastal Plain; Gulf of California

mica deposits
Term introduced in 1981. Includes use on level 1 and 2 as a commodity term (list C).
—— CR see also under economic geology under ——

mica group
Includes use in combination with sheet silicates (i.e. sheet silicates, mica group) on level 2 under minerals(1). See list L.

Michigan
Includes use on level 1 as an area term (list O). For term set options see list B.

microcraters
Includes use on level 2 under Moon(1).

microearthquakes
Includes use on level 2 under seismology(1).

microfossils see problematic microfossils

Micronesia
Collective name for islands E of the Philippine Islands and S of Japan. Index island groups as applicable. Includes use on level 1 or 2 as an area term (list O). If 1, see term set options under list B.
—— CR see also Mariana Islands; Marshall Islands

micropaleontology
Used for the discipline as a whole, not for specific applications. Includes use on level 1 (list A). See list F (microfossil groups). Term set options are:
topic [applications, bibliography, catalogs, concepts, experimental studies, history, instruments, methods, photography, practice, techniques, theoretical studies]
subtopic [e.g. sample preparation]
—— CR see also palynology

microscopy see electron microscopy

microseisms
Includes use on level 2 under seismology(1); on level 3 under earthquakes(1).

mid-ocean ridges
Includes use on level 2 under ocean floors(1).
—— CR see under ocean floors

Middle East
An indefinite and unofficial term comprising a region including Cyprus and the countries of SW Asia. Index countries as applicable. Includes use on level 1 or 2 as an area term (list O). If 1, see term set options under list B.
—— CR see also Cyprus; Iraq; Israel; Jordan; Lebanon; Syria; Turkey

Midwest
A level 1 area term as of 1978. N part of central U.S. Region comprising states N of the Ohio and Missouri Rivers plus the E edge of the Great Plains including Kansas and Nebraska.

migmatites
Includes use on level 2 under metamorphic rocks(1). See list J.

migration of elements
Includes use on level 2 under metamorphism(1); on level 3 under geochemistry(1) and under weathering(1).

Miliolacea
Includes use on level 2 under foraminifera(1). See list F.

Miliolina
Includes use on level 2 under foraminifera(1). See list F.

mineral assemblages
Includes use on level 2 under metamorphic rocks(1).

mineral data
Includes use on level 2 under clay mineralogy(1); on level 3 under minerals(1).

mineral deposits
Do not include water resources or fuels. Used for substantial discussions of the genesis of ore deposits. Includes use on level 1 (list A) in combination with genesis (i.e. mineral deposits, genesis). Term set options are:
commodity (List C)
topic
controls
type of control [geochemical controls, hydrogeological controls, lithologic controls, mechanical controls, paleogeographic controls, stratigraphic controls, structural controls]
processes
type of process [endogene processes, epigene processes, exhalative processes, exogene processes, hydrothermal processes, igneous processes, metamorphism, plate tectonics, sedimentary processes, supergene processes, syngenesis, volcanism, weathering]
topic [age, causes, concepts, environment, experimental studies, interpretation, mechanism, patterns, theoretical studies]
subtopic

mineral exploration
Used for substantial discussions of exploration for "commodities" (list C) excluding fuel deposits and water resources. Used primarily for the application of the methods listed below. Includes use on level 1 (list A). Term set options are:

methods [biogeochemical methods, geobotanical methods, geochemical methods, geological methods, geomorphological methods, geophysical methods, hydrological methods, methods, photogeologic methods, remote sensing]
topic
topic [applications, bibliography, concepts, history, instruments, objectives, ore guides, programs, techniques]
subtopic

mineral inclusions
Includes use on level 2 under inclusions(1).
—— CR see under inclusions

mineral prospecting CR see mineral exploration

mineral resources
For very general treatments. Includes use on level 1 as a commodity term (list C); on level 2 under ocean floors(1).
—— CR see also under economic geology under ——

mineral waters
Includes use on level 2 and 3 under springs(1).

mineralogy
Used for the discipline as a whole. Includes use on level 1 (list A); on level 2 under areas (list B), under symposia(1), and under education(1). If 1, term set options are:
topic [applications, bibliography, catalogs, classification, concepts, education, experimental studies, history, instruments, methods, nomenclature, objectives, philosophy, practice, principles, symposia, textbooks, theoretical studies]
subtopic
—— CR see also crystallography
—— see clay mineralogy; optical mineralogy

minerals
Includes use on level 1 (list A); on level 2 under phase equilibria(1) and weathering(1). Used for descriptions of mineral occurrence or of minerals themselves. If 1, term set options are:
mineral group (List L)
mineral species
topic [chemical analysis, crystal chemistry, crystal structure, mineral data, occurrence, optical properties, spectra] (more than one single mineral)
subtopic
—— CR see also crystal chemistry; crystal growth; crystal structure
—— see clay minerals; heavy minerals; industrial minerals; miscellaneous minerals; silica minerals

mines
As of 1978, term is used on level 2 under land subsidence(1), land use(1), and underground installations(1).

mining geology
Geology applied to mining operations. Includes use on level 1 (list A). Term set options are:
topic [applications, bibliography, catalogs, classification, concepts, evaluation, history, instruments, methods, nomenclature, objectives, practice, production control, symposia, technology, textbooks]
subtopic

Minnesota
Includes use on level 1 as an area term (list O). For term set options see list B.

Miocene
World. Above Oligocene, below Pliocene. Includes use on level 1 as an age term (list E); on level 2 under paleo- terms, e.g. paleoecology, paleogeography, paleomagnetism; on level 3 under age terms(1).

——— CR see also under geochronology under ———; see also under stratigraphy under ———

miospores
Includes use on level 2 under palynomorphs(1). See list F.

miscellanea
Includes use on level 2 under fossil group(1) if fossil is not yet classified. See list F.

miscellaneous minerals
Includes use on level 2 under minerals(1) for the case of several mineral groups when the topic is not emphasized or for a mineral of unknown affinity. See list L.

Mississippi
Includes use on level 1 as an area term (list O). For term set options see list B.

Mississippi Valley
A level 1 area term as of 1978. River valley. Index states as applicable.

Mississippian
Includes use on level 1 as an age term (list E); on level 2 under paleoterms, e.g. paleoecology, paleogeography, paleomagnetism. After the Devonian and before the Pennsylvanian. Approximate equivalent of Lower Carboniferous of European usage.

——— CR see also Carboniferous; see also under geochronology under ———; see also under stratigraphy under ———

Missouri
Includes use on level 1 as an area term (list O). For term set options see list B. y

models
Includes use on level 2 under ground water(1).

Moeritherioidea
Term introduced in 1981. Includes use on level 2 under Mammalia(1). See list F.

Mohorovicic discontinuity
Includes use on level 1 (list A) for discussions which stress the crust-mantle transition. Term set options are:
topic [causes, composition, depth, detection, geometry, identification, interpretation, Mohole project, observations, patterns, properties, temperature]
subtopic (no area term)

——— CR see also crust; mantle; see also under seismology under ———; see also under tectonophysics under ———

Mollusca
Includes use on level 1 and 2 as a fossil term (list F).

mollusks
As of 1981, restricted to use as the common name for Mollusca. See list F. To be used for Mollusca when they are used as tools in stratigraphy. Includes use on level 1. If 1, term set options are:
biostratigraphy
age [List E]

molybdates
Includes use on level 2 under minerals(1). See list L.

——— CR see under minerals

molybdenum
Includes use on level 1 and 2 as a chemical element (list D). As of 1981, use molybdenum ores for molybdenum as a commodity. Before 1981, included use on level 1 as a commodity term.

molybdenum ores
Term introduced in 1981. Includes use on level 1 and 2 as a commodity term (list C).

——— CR see also under economic geology under ———

Monaco
On the Mediterranean Sea near the French-Italian border. Includes use on level 1 as an area term (list O). For term set options see list B.

monazite deposits
Term introduced in 1981. Includes use on level 1 and 2 as a commodity term (list C).

——— CR see also under economic geology under ———

Mongolia
Includes use on level 1 as an area term (list O). For term set options see list B.

Monocotyledoneae
Includes use on level 2 under angiosperms(1). See list F.

Monograptina
Includes use on level 2 under Graptolithina(1). See list F.

Monoplacophora
Includes use on level 2 under Mollusca(1). See list F.

Monotremata
Includes use on level 2 under Mammalia(1). See list F.

Montana
Includes use on level 1 as an area term (list O). For term set options see list B.

monzonites
Term introduced in 1981. Includes use on level 2 under igneous rocks(1). See list H.

——— CR see under igneous rocks

Moon
See also other planets, i.e. Mars, Venus, etc. Includes use on level 1. Term set options are:
name of discipline [List B except for oceanography, paleobotany and paleontology]
topic (similar to terms called for on 3rd level in List B)
theoretical studies
subtopic
topic [age, anomalies, atmosphere, bibliography, catalogs, composition, concepts, environment, evolution, exploration, genesis, gravity field, lunar crust, lunar interior, magnetic field, microcraters, motions, observations, paleomagnetism, planetary interiors, satellites, surface properties]
subtopic

Morocco
Includes use on level 1 as an area term (list O). For term set options see list B.

morphology
Includes use on level 2 under fossil terms (list F), and under aurora(1) and soils(1).

motions
Includes use on level 2 under Earth(1) and Moon(1); on level 3 under aurora(1) morphology(2).

movement
Includes use on level 2 under ground water(1) and plate tectonics(1); on level 3 under soils(1) water regimes(2).

movements see mass movements

Mozambique
Includes use on level 1 as an area term (list O). For term set options see list B.

mud volcanoes
Includes use on level 1 (list A). Used for an accumulation, usually conical,

of mud and rock ejected by volcanic gases; also for a similar accumulation formed by escaping petroliferous gases. Term set options are: topic [age, causes, classification, detection, distribution, evolution, genesis, interpretation, mechanism, occurrence, temperature] subtopic (no area term)

—— CR see also volcanology; see also under geomorphology under ____; see also under sedimentary petrology under ____

mudflows
As of 1978, term is used on level 2 under geologic hazards(1), and slope stability(1).

mullions CR see under ____ under lineation

Multituberculata
Includes use on level 2 under Mammalia(1). See list F.

Musci
Includes use on level 2 under bryophytes(1). See list F.

museums
Includes use on level 1 (list A). Used for papers discussing the functioning of a geology museum and/or its history. Term set options are:
 topic (list B)
 name of museum (in original language)

—— CR see also under general under ____

mylonites
Includes use on level 2 under metamorphic rocks(1). See list J.

Myodocopina
Term introduced in 1981. Includes use on level 2 under Ostracoda(1). See list F.

Myomorpha
Term introduced in 1981. Includes use on level 2 under Mammalia(1). See list F.

Myriapoda
Includes Archipolypoda, Chilopoda, Diplopa, Pauropoda, and Symphyla. Includes use on level 2 under Arthropoda(1).

Myxomycetes
Term introduced in 1981. Includes use on level 2 under fungi(1). See list F.

Myzostomia
As of 1981, includes use on level 2 under worms(1). See list F. Before 1981, included use on level 2 under Annelida(1).

Namibia
Term introduced in 1981 to replace South-West Africa. Includes use on level 1 as an area term (list O). For term set options see list B.

nannofossils
Includes discoasters and Nannoconus. Includes use on level 2 under algae(1).

nappes CR see under ____ under folds; tectonics

Nassellina
Suborder. Includes use on level 2 under Radiolaria(1). See list F.

native elements
Term introduced in 1981. Includes use on level 2 under minerals(1). See list L.

—— CR see under minerals

natural gas
Not a valid term through 1977. After 1977, includes use on level 1 as a commodity term.

—— CR see also under economic geology under ____

natural resources
As of 1978, term is used on level 2 under conservation(1), land use(1), and reclamation(1).

—— CR see under conservation

Nautiloidea
As of 1981, includes use on level 2 under Mollusca(1). See list F. Before 1981, included on level 3 under Mollusca(1) Cephalopoda(2).

Nebraska
Includes use on level 1 as an area term (list O). For term set options see list B.

Nematoida
Includes use on level 2 under worms(1). See list F.

Nematomorpha
Includes use on level 2 under worms(1). See list F.

Nemerta
Includes use on level 2 under worms(1). See list F.

neodymium
Includes use on level 1 and 2 as a chemical element (list D).

Neogastropoda
Term introduced in 1981. Includes use on level 2 under Mollusca(1). See list F.

Neogene
As of 1978, term includes use on level 1 and 2 (list E). An interval of geologic time incorporating the Miocene and Pliocene of the Tertiary period; the upper Tertiary.

—— CR see also under geochronology under ____; see also under stratigraphy under ____

neon
Includes use on level 1 and 2 as a chemical element (list D).

Neornithes
Includes use on level 2 under Aves(1). See list F.

neotectonics
A level 1 term as of 1978. Used for the study of the last structures and structural history of the Earth's crust, after the Miocene and during the later Tertiary and the Quaternary. Term set options are:
 concepts
  subtopic
 effects
  topic [e.g. changes of level, geomorphologic effects, seismicity]
 measurement
  subtopic
 observations
  subtopic
 rates
  subtopic [e.g. absolute age, geodetic coordinates, remote sensing, satellite measurements]
 subsidence
  subtopic
 uplifts
  subtopic

—— CR see also changes of level; isostasy; tectonics; see also under structural geology under ____; see also under tectonophysics under ____

Nepal
Includes use on level 1 as an area term (list O). For term set options see list B.

nepheline group
Includes use in combination with framework silicates (i.e. framework silicates, nepheline group) on level 2 under minerals(1). See list L.

Neptune
Includes use on level 1 and 2. See entry under Moon(1) for term set options.

neptunium
Includes use on level 1 and 2 as chemical element (list D).

nesosilicates
Term introduced in 1981. Includes use on level 2 under minerals(1). See list L.

Netherlands
Includes use on level 1 as an area term (list O). For term set options see list B.

Netherlands Antilles
Formerly known as Curacao territory. Islands of Aruba, Bonaire, and Curacao in the Caribbean Sea off the coast of Venezuela, plus the Dutch section of St. Martin at N end of Leeward Islands. Includes use on level 1 as an area term (list O). For term set options see list B.

Neuropteroida
Term introduced in 1981. Includes use on level 2 under Insecta(1). See list F.

Nevada
Includes use on level 1 as an area term (list O). For term set options see list B.

New Brunswick
Includes use on level 1 as an area term (list O). For term set options see list B.

New Caledonia
French overseas territory E of Queensland, Australia. Also its main island. Includes use on level 1 as an area term (list O). For term set options see list B.

New England
Index states as applicable. Includes use on level 1 as an area term (list O). For term set options see list B.

New Guinea
Refers to whole island. Index Irian Jaya and/or Papua New Guinea. Includes use on level 1 as an area term (list O). For term set options see list B.

New Hampshire
Includes use on level 1 as an area term (list O). For term set options see list B.

New Hebrides
Group of islands NE of New Caledonia and W of Fiji. Under joint British and French administration. Includes use on level 1 as an area term (list O). For term set options see list B.

New Jersey
Includes use on level 1 as an area term (list O). For term set options see list B.

New Mexico
Includes use on level 1 as an area term (list O). For term set options see list B.

New South Wales
Includes use on level 1 as an area term (list O). For term set options see list B.

New York
Includes use on level 1 as an area term (list O). For term set options see list B.

New Zealand
Includes use on level 1 as an area term (list O). For term set options see list B.

Newfoundland
As of 1981, refers to the island of Newfoundland. Before 1981, referred to the province, which includes the island of Newfoundland plus Labrador. Includes use on level 1 as an area term (list O). For term set options see List B.

Nicaragua
Includes use on level 1 as an area term (list O). For term set options see list B.

nickel
Includes use on level 1 and 2 as a chemical element (list D). As of 1981, use nickel ores for nickel as a commodity. Before 1981, included use on level 1 as a commodity term.

nickel ores
Term introduced in 1981. Includes use on level 1 and 2 as a commodity term (list C).

—— *CR see also under* economic geology *under* ——

Niger
Includes use on level 1 as an area term (list O). For term set options see list B.

Nigeria
Includes use on level 1 as an area term (list O). For term set options see list B.

Nilssoniales
Includes use on level 2 under gymnosperms(1). See list F.

niobates
As of 1981, includes use on level 2 under minerals(1). See list L. Before 1981, included use on level 3 under minerals(1) oxides(2).

—— *CR see under* oxides *under* minerals

niobium
Includes use on level 1 and 2 as a chemical element (list D). As of 1981, use niobium ores for niobium as a commodity. Before 1981, included use on level 1 as a commodity term.

niobium ores
Term introduced in 1981. Includes use on level 1 and 2 as a commodity term (list C).

—— *CR see also under* economic geology *under* ——

niobotantalates
Term introduced in 1981. Includes use on level 2 under minerals(1). See list L.

nitrate deposits
Term introduced in 1981. Includes use on level 1 and 2 as a commodity term (list C).

—— *CR see also under* economic geology *under* ——

nitrates
Includes use on level 2 under minerals(1). See list L. As of 1981, use nitrate deposits for nitrate as a commodity. Before 1981, included use on level 1 as a commodity term.

—— *CR see under* minerals

nitrides
As of 1981, includes use on level 2 under minerals(1). See list L. Before 1981, included use on level 3 under minerals(1) native elements and alloys(2).

—— *CR see under* native elements and alloys *under* minerals

nitrogen
Includes use on level 1 and 2 as a chemical element (list D); on level 3 under soils(1).

noble gases
Includes use on level 1 and 2. See list D for term set options. Also see list C (commodities).

—— *CR see also* argon; helium; krypton; neon; radon; xenon

Nodosariacea
Includes use on level 2 under foraminifera(1). See list F.

nodules
Includes use on level 1 (list A). As of 1981, restricted to nodules on the ocean floor. For nodules in kimberlite, use xenoliths. For nodules within sedimentary formations such as chert nodules, use concretions. If level 1, term set options are:
kind of nodule [ferromanganese composition, manganese composition]
topic
topic [age, classification, composition, distribution, genesis, observations, properties]
subtopic

Noeggerathiales
Includes use on level 2 under pteridophytes(1). See list F.

Nomarthra
Term introduced in 1981. Includes use on level 2 under Mammalia(1). See list F.

nomenclature
Includes use as a level 2 or 3 term appropriate to a large number of topics, e.g. on level 2 under sedimentary petrology(1). See list G.

nonmetal deposits
Term introduced in 1981. Includes use on level 1 and 2 as a commodity term (list C).

—— *CR see also under* economic geology *under* ——

nonmetals
Includes use on level 1 and 2 as a chemical element (list D). As of 1981, use nonmetal deposits for nonmetals as a commodity. Before 1981, included use on level 1 as a commodity term.

—— *CR see also* boron deposits; fluorspar; iodine; phosphate deposits; potash deposits; sulfur deposits

North America
Includes use on level 1 or 2 as an area term (list O). If 1, see term set options under list B. Include Canada, United States, Mexico, and Saint Pierre and Miquelon.

—— *CR see also* Appalachians; Atlantic Coastal Plain; Canada; Great

Lakes; Great Lakes region; Great Plains; Gulf Coastal Plain; Mexico; Rocky Mountains; United States

North Carolina
Includes use on level 1 as an area term (list O). For term set options see list B.

North Dakota
Includes use on level 1 as an area term (list O). For term set options see list B.

North Sea
Between the European continent on the S and E, and Great Britain on the W. Includes use on level 1 as an area term (list O). For term set options see list B.

Northern Hemisphere
Used when discussing many large areas too numerous to mention. Includes use on level 1 and 2 as an area term (list O). If 1, see term set options under list B.

—— CR see also Africa; Arctic Ocean; Asia; Atlantic Ocean; Central America; Eurasia; Europe; North America; Pacific Ocean; USSR

Northern Ireland
Six counties comprising the NE part of island of Ireland. Includes use on level 1 as an area term (list O). For term set options see list B.

Northern Territory
Includes use on level 1 as an area term (list O). For term set options see list B.

Northwest Territories
Index districts as applicable. Includes use on level 1 as an area term (list O). For term set options see list B.

Norway
Includes use on level 1 as an area term (list O). For term set options see list B.

Nothosauria
Term introduced in 1981. Includes use on level 2 under Reptilia(1). See list F.

Notoungulata
Order. Includes use on level 2 under Mammalia(1). See list F.

Nova Scotia
Includes use on level 1 as an area term (list O). For term set options see list B.

nuclear explosions
Includes use on level 3 under seismology(1) explosions(2); on level 2 under explosions(1).

—— CR see under explosions; see under explosions under seismology

nuclear facilities
A level 1 term as of 1978. Before 1978, included use on level 2 under engineering geology(1). If 1, term set options are:

design
  subtopic
earthquakes
  subtopic
faults
  subtopic
feasibility studies
  subtopic
foundations
  subtopic
geologic hazards
  subtopic
impact statements
  subtopic
pollution
  subtopic
rock mechanics
  subtopic
seepage
  subtopic
site exploration
  subtopic
soil mechanics
  subtopic
—— CR see also under engineering geology under ____

Nummulitidae
Includes use on level 2 under foraminifera(1). See list F.

nutrients
Includes use on level 2 under soils(1). See list M.

obduction
Includes use on level 2 under plate tectonics(1).

objectives
Includes use as a level 2 or 3 term appropriate to a large number of topics, e.g. on level 2 under geology(1). See list G.

observations
Includes use as a level 2 or 3 term appropriate to a large number of topics, e.g. on level 2 under seismology(1). See list G.

observatories
Includes use on level 2 under geophysics(1) and seismology(1).

occurrence
Includes use as a level 2 or 3 term appropriate to a large number of topics, e.g. on level 2 under metamorphic rocks(1). See list G.

ocean see circulation in the ocean

ocean basins
Used for studies on the major ocean basins, their origin, evolution and present configuration. For basins found within the oceans and for sedimentation studies, see ocean floors. Includes use on level 1 (list A). Term set options are:
topic [age, evolution, genesis, patterns]
  subtopic (no area term)
—— CR see also under oceanography under ____; see also under tectonophysics under ____

ocean circulation
Level 1 term introduced in 1978. Primarily used for special indexes. Term set options are:
currents
  (name of current)
topic [anomalies, biocirculation, boundary layer, causes, climate-induced circulation, Coriolis force, detection, diffusion, distribution, Ekman spiral, genesis, patterns, thermal circulation, thermohaline circulation, tides, turbulence]
  subtopic
—— CR see also under oceanography under ____

ocean floors
Used for discussions of processes taking place on the ocean floor as well as features thereof. For tectonics, see ocean basins. Includes use on level 1 (list A). Term set options are:
topic [anomalies, bottom features, exploration, mid-ocean ridges, mineral resources, seamounts, sedimentation, submarine canyons, trenches, troughs]
  subtopic (no area term)
—— CR see also under oceanography under ____; see also under tectonophysics under ____

ocean waves
Includes use on level 1 (list A) for special indexes or for papers relating physical oceanography to marine geology. Term set options are:
topic [anomalies, breaking waves, catastrophic waves, causes, effects, genesis, ideal waves, internal waves, shoaling, transformations]
  subtopic ·
—— CR see also under oceanography under ____

Oceania
Collective name for the islands and island groups of the central and south Pacific Ocean. Index island divisions as applicable: Melanesia, Micronesia, and Polynesia. Includes use on level 1 or 2 as an area term (list O). If 1, see term set options under list B.

oceanography
Includes use on level 1 (list A) for special indexes and for treating the discipline as a whole; on level 2 under area terms (list B), bibliography(1), education(1), continental shelf(1), and continental slope(1). In January 1976, the following terms were deleted from level 2: boundary layer, diffusion, distribution, and turbulence. These terms are now used on level 2 under ocean circulation(1). When oceanography is a level 1 term, set options are:

experimental studies
  topic
instruments
  name of instrument or platform
phenomenon
sea ice
  topic
techniques
  topic
theoretical studies
  topic
topic [applications, bibliography, catalogs, classification, concepts, education, general, history, methods, nomenclature, objectives, practice, principles, research, symposia, textbooks]
  subtopic

Octocorallia
Includes use on level 2 under Coelenterata(1). See list F.

Odonatopteroida
Term introduced in 1981. Includes use on level 2 under Insecta(1). See list F.

Odontopleurida
Includes use on level 2 under Trilobita(1). See list F.

Ohio
Includes use on level 1 as an area term (list O). For term set options see list B.

oil CR see petroleum

oil and gas fields
For detailed descriptions of individual fields or for discussions of the origin of several fields. Includes use on level 1 (list A). See also list C (commodities). Term set options are:
  topic [classification, distribution, genesis]
  subtopic (no area term)
—— CR see also natural gas; petroleum; see also under economic geology under ——

oil sands
Includes use on level 1 as a commodity term (list C). As of 1981, includes use on level 3 under sedimentary rocks(1) organic residues(2). See list I.
—— CR see also under economic geology under ——

oil shale
Includes use on level 1 as a commodity term (list C).
—— CR see also under economic geology under ——

oil spills
As of 1978, term is used on level 2 under pollution(1).
—— CR see under pollution

Okhotsk Sea
West of Kamchatka Peninsula and the Kuril Islands. Includes use on level 1 as an area term (list O). For term set options see list B.

Oklahoma
Includes use on level 1 as an area term (list O). For term set options see list B.

Oligocene
World. Above Eocene, below Miocene. Includes use on level 1 as an age term (list E); on level 2 under paleo- terms, e.g. paleoecology, paleogeography, paleomagnetism.
—— CR see also under geochronology under ——; see also under stratigraphy under ——

Oligochaetia
As of 1981, includes use on level 2 under worms(1). See list F. Before 1981, included use on level 2 under Annelida(1).

olistostromes CR see under —— under sedimentary structures; tectonics

olivine group
Includes use in combination with orthosilicates (i.e. orthosilicates, olivine group) on level 2 under minerals(1). See list L.

Oman
Formerly Muscat and Oman. Includes use on level 1 as an area term (list O). For term set options see list B.

Ontario
Includes use on level 1 as an area term (list O). For term set options see list B.

ontogeny
Includes use on level 2 and 3 under fossil group(1). See list F.

ooze CR see under —— under sediments

Ophiocistioidea
Includes use on level 2 under Echinodermata(1). See list F.

ophiolite CR see under —— under igneous rocks; metamorphic rocks; plate tectonics; tectonics

Ophiuroidea
As of 1981, includes use on level 2 under Echinodermata(1). Before 1981, included use on level 3 under Echinodermata(1) Stelleroidea(2). See list F.

optical mineralogy
Includes use on level 2 under geochronology(1); on level 3 under minerals(1) methods(2).

optical properties
As of 1978, term is used on level 2 under minerals(1).

Orbitoidacea
Includes use on level 2 under foraminifera(1). See list F.

Orbitoididae
Includes use on level 2 under foraminifera(1). See list F.

order-disorder
As of 1978, term is used on level 2 under crystal chemistry(1).

Ordovician
Includes use on level 1 as an age term (list E); on level 2 under paleo- terms, e.g. paleoecology, paleogeography, paleomagnetism. After the Cambrian, before the Silurian.
—— CR see also under geochronology under ——; see also under stratigraphy under ——

ore see genesis of ore deposits

ore guides
Includes use on level 2 under mineral exploration(1).
—— CR see under mineral exploration

Oregon
Includes use on level 1 as an area term (list O). For term set options see list B.

ores see aluminum ores; antimony ores; arsenic ores; beryllium ores; bismuth ores; cadmium ores; cerium ores; chromite ores; cobalt ores; copper ores; gold ores; iron ores; lead ores; lithium ores; magnesium ores; manganese ores; mercury ores; metal ores; molybdenum ores; nickel ores; niobium ores; platinum ores; polymetallic ores; pyrite ores; silver ores; strontium ores; tantalum ores; thorium ores; tin ores; titanium ores; tungsten ores; uranium ores; vanadium ores; zinc ores

organic compounds
Includes use on level 2 under minerals(1). See list L.
—— CR see under minerals

organic materials
Includes use on level 1 (list A); on level 2 under soils(1). Used for discussions of mostly very small concentrations of organic materials in rocks. If 1, term set options are:
  kind of material [amino acids, bitumens, carbohydrates, fatty acids, humates, humic acids, hydrocarbons, kerogen, phenols]
  specific kind of material or topic
  topic [abundance, age, alteration, analysis, classification, composition, detection, distribution, experimental studies, genesis, geochemistry, identification, nomenclature, observations, occurrence, properties, varieties]
  subtopic
—— CR see also bitumens; lignite; natural gas; oil sands; oil shale; peat; petroleum

organic residues
As of 1981, includes use on level 2 under sedimentary rocks(1) and sediments(1). See lists I and N. Before 1981, included use on level

2 under sedimentary rocks(1). Before 1976, included use on level 2 under sedimentary rocks(1) and sediments(1).

organization
As of 1981, includes use on level 2 under survey organizations(1). Before 1981, included use on level 2 under surveys(1).

orientation
Includes use as attitude of fold elements with respect to external coordinates. Includes use on level 2 under faults(1) and folds(1); on level 3 to index orientation of grains under structural analysis(1).

—— see preferred orientation

origin see life origin

Ornithischia
As of 1981, includes use on level 2 under Reptilia(1). See list F. Before 1981, included use on level 3 under Reptilia(1) Archosauria(2).

orogeny
Includes use on level 1 (list A). Used for discussions of either individual orogenies or detailed general treatments on several orogenies. Term set options are:
  topic [absolute age, causes, evolution, mechanism, periodicity]
  name of orogeny
  subtopic

—— CR see also epeirogeny; geosynclines; tectonics

Orthida
Includes use on level 2 under Brachiopoda(1). See list F.

orthoamphibole
Term introduced in 1981. Includes use in combination with chain silicates (i.e. chain silicates, orthoamphibole) on level 2 under minerals(1). See list L.

Orthopteroida
Term introduced in 1981. Includes use on level 2 under Insecta(1). See list F.

orthopyroxene
As of 1981, includes use in combination with chain silicates (i.e. chain silicates, orthopyroxene) on level 2 under minerals(1). See list L. Before 1981, included use on level 3 under minerals(1) chain silicates, pyroxene group(2).

orthosilicates
Sorosilicates and nesosilicates. Includes use on level 2 under minerals(1); in combination with epidote group, garnet group, humite group, melilite group, and olivine group (i.e. orthosilicates, epidote group) to form terms on level 2 under minerals(1). See list L.

—— CR see under minerals

osmium
Includes use on level 1 and 2 as a chemical element (list D).

Osteichthyes
Includes use on level 2 under Pisces(1). See list F.

Ostracoda
Includes use on level 1 and 2 as a fossil term (list F).

ostracods
Term introduced in 1981 as the common name for Ostracoda. See list F. To be used for Ostracoda when they are used as tools in stratigraphy. Includes use on level 1. If 1, term set options are:
  biostratigraphy
  age [List E]

Ostreacea
Term introduced in 1981. Includes use on level 2 under Mollusca(1). See list F.

oxalates
Term introduced in 1981. Includes use on level 2 under minerals(1). See list L.

oxides
Includes niobates and tantalates. Includes use on level 2 under minerals(1). See list L.

—— CR see under minerals

oxygen
Includes use on level 1 and 2 as a chemical element (list D).

oxysulfides
As of 1981, includes use on level 2 under minerals(1). See list L. Before 1981, included use on level 3 under minerals(1) sulfides(2).

—— CR see under sulfides under minerals

P-T conditions
Includes use on level 2 under metamorphism(1). As of 1981, includes use on level 2 under inclusions(1) and fluid inclusions(1).

Pacific Coast
Region comprising those states fronting on the Pacific Ocean. Index states as applicable. Introduced as level 1 area term in 1978.

Pacific Ocean
Includes use on level 1 or 2 as an area term (list O). If 1, see term set options under list B.

—— CR see also Bering Sea; Celebes Sea; China Sea; Coral Sea; East Tasman Sea; Galapagos Islands; Gulf of California; Japan Sea; Melanesia; Micronesia; New Caledonia; Oceania; Okhotsk Sea; Philippine Sea; Polynesia; South China Sea; Yellow Sea

Pacific region
Includes use on level 1 or 2 as an area term (list O). If 1, see term set options under list B.

Pakistan
Formerly consisted of East Pakistan and West Pakistan which were separated by about 1,000 miles of Indian territory. After East Pakistan became the independent state of Bangladesh in 1971, West Pakistan and Pakistan became coextensive. Includes use on level 1 as an area term (list O). For term set options see list B.

Palaeodictyopteroida
Term introduced in 1981. Includes use on level 2 under Insecta(1). See list F.

paleobotany
For general discussion of fossil plants. See names of major floral groups (list F). Includes use on level 1 (list A); on level 2 under area terms (list B). If level 1, term set options are:
  topic [applications, bibliography, catalogs, classification, concepts, education, evolution, history, instruments, methods, nomenclature, objectives, practice, principles, symposia, textbooks]
  subtopic

Paleocene
World. Lower Tertiary, above Gulfian (Cretaceous), below Eocene. Includes use on level 1 as an age term (list E); on level 2 under paleoterms, e.g. paleoecology, paleogeography; on level 3 under age terms(1).

—— CR see also under geochronology under ____; see also under stratigraphy under ____

paleoclimatology
Includes use on level 1 (list A). Used for treatments of the climate of a given period of time in the geologic past. If 1, term set options are:
  age (single term from list E)
  (area)
  topic [applications, changes, concepts, cycles, evolution, indicators, interpretation, methods, paleotemperature, patterns]
  subtopic

paleoecology
Includes use on level 1 (list A); on level 2 under fossil terms (list F). Used for the study of the relationships between organisms and their environments, the death of organisms, and their burial and postburial history in the geologic past based on fossil fauna and flora and their stratigraphic position. If 1, term set options are:
  age [List E]
  area
  fossil group [List F]

age [List E]
topic [analysis, changes, indicators, interpretation, sedimentation]
subtopic [e.g. name of environment: deltaic environment, estuarine environment, fluvial environment, lacustrine environment, lagoonal environment, marine environment, paludal environment, reefs, terrestrial environment]

Paleogene
As of 1978, includes use on level 1 and 2 (list E). An interval of geologic time incorporating the Oligocene, Eocene, and Paleocene of the Tertiary; the lower Tertiary.
—— *CR see also under* geochronology *under* ——; *see also under* stratigraphy *under* ——

paleogeography
Includes use on level 1 (list A). Used for the geography of ancient times, specifically the study and description of the physical geography of the geologic past. Term set options are:
age [List E; a single term]
(area)
topic [applications, changes, concepts, interpretation, maps, methods, patterns, principles]
subtopic (no area term)
—— *CR see also under* stratigraphy *under* ——

paleomagnetism
Used for the study of natural remanent magnetization. Includes use on level 1 (list A); on level 2 under Moon(1) and geochronology(1); on level 3 under plate tectonics(1). If 1, term set options are:
age [List E]
(area)
topic [applications, causes, changes, concepts, experimental studies, geochemistry, interpretation, methods, patterns, polar wandering, pole positions, reversals, stability]
subtopic

paleontology
Used for discipline as a whole. Includes use on level 1 (list A); on level 2 under area terms (list B), age terms (list E), continental drift(1), symposia(1), bibliography(1), and automatic data processing(1). See also names of fossil groups (list F). If 1, term set options are:
topic [applications, bibliography, catalogs, classification, concepts, education, evolution (as a concept), fossilization, history, instruments, life origin, nomenclature, practice, principles, symposia, taxonomy (principles of), textbooks]
subtopic

paleosalinity
As of 1978, term is used on level 2 under fluid inclusions(1).

Paleosols *CR see under* —— *under* ——

paleotemperature
Term reintroduced in 1982. A valid term through 1973. Includes use on level 2 under paleoclimatology(1).
—— *CR see* geologic thermometry *under* fluid inclusions

Paleozoic
Includes use on level 1 as an age term (list E); on level 2 under paleoterms, e.g. paleoecology, paleogeography, paleomagnetism.
—— *CR see also* Cambrian; Carboniferous; Devonian; Mississippian; Ordovician; Pennsylvanian; Permian; Silurian; *see also under* geochronology *under* ——; *see also under* stratigraphy *under* ——

palladium
Includes use on level 1 and 2 as a chemical element (list D).

palynology
Includes use on level 1 (list A). See list F. Used for the study of pollen of seed plants and spores of other embryophytic plants whether living or fossil, including their dispersal and applications. Term set options are:
topic [applications, bibliography, catalogs, classification, concepts, education, fossilization, history, instruments, methods, nomenclature, practice, principles, symposia, techniques, textbooks]
subtopic

palynomorphs
Includes use on level 1 and 2 as a fossil term (list F).

Panama
Includes use on level 1 as an area term (list O). For term set options see list B.

Pangaea
Name proposed for the supercontinent comprising all the landmasses of Earth which existed about 300 million years ago prior to continental drift. Includes use on level 2 under continental drift(1).

Pantodonta
Term introduced in 1981. Includes use on level 2 under Mammalia(1). See list F.

Pantotheria
Includes use on level 2 under Mammalia(1). See list F.

Papua New Guinea
Eastern half of the island of New Guinea. Comprises Papua and the former Australian U.N. trusteeship of The Territory of New Guinea plus the Bismarck Archipelago, and Bougainville, Buka and Green islands of

the W Solomon Islands. Became an independent state on September 16, 1976. Includes use on level 1 as an area term (list O).

Parablastoidea
Includes use on level 2 under Echinodermata(1). See list F.

Paracrinoidea
Includes use on level 2 under Echinodermata(1). See list F.

paragenesis
For detailed treatment of mineral sequences in metamorphosed or altered rocks and mineral deposits. Includes use on level 1 (list A). Term set options are:
kind of mineral deposit (first level term only)
area
kind of rock (first level term only)
area
topic [changes, evolution, interpretation, observations, patterns, processes, rates]
subtopic

Paraguay
Includes use on level 1 as an area term (list O). For term set options see list B.

Pareiasauria
Term introduced in 1981. Includes use on level 2 under Reptilia(1). See list F.

particle precipitation
Includes use on level 2 under ionosphere(1).

particle radiation
Includes use on level 2 under astrophysics and solar physics(1).

particle-track dating
Cosmic-ray tracks. Includes use on level 2 under geochronology(1).
—— *CR see under* geochronology

particles
General term. Includes use on level 2 and 3 under meteorology(1); on level 3 under interplanetary space(1) cosmic rays(2).
—— *see* trapped particles

partitioning
As of 1978, term is used on level 2 under crystal chemistry(1).

patterned ground *CR see under* periglacial geology *under* glacial geology

patterns
Includes use as a level 2 or 3 term appropriate to a large number of topics, e.g. on level 2 under faults(1), fractures(1), and under soils(1). See list G.
—— *see* regional patterns

Paucituberculata
Term introduced in 1981. Includes use on level 2 under Mammalia(1). See list F.

peat
Includes use on level 1 as a commodity term (list C). As of 1981, includes use on level 3 under sediments(1) organic residues(2). See list N. Before 1981, included use on level 3 under sediments(1) organic sediments(2).

—— CR see also under —— under sediments; see also under economic geology under ——

Pectinacea
As of 1981, includes use on level 2 under Mollusca(1). See list F. Before 1981, included use on level 3 under Mollusca(1) Bivalvia(2).

pedogenesis
Term introduced in 1982. Includes use on level 2 under soils(1).

pegmatite
As of 1981, includes use on level 3 under igneous rocks(1) granites(2). See list H. Before 1981, included use on level 3 under igneous rocks(1) granite-granodiorite family(2). Also includes use on level 1 as a commodity term (list C).

—— CR see also under —— under igneous rocks; see also under economic geology under ——

Pelecypoda CR see Bivalvia under Mollusca

Pelycosauria
Term introduced in 1981. Includes use on level 2 under Reptilia(1). See list F.

peneplains CR see under erosion features under geomorphology

Pennsylvania
Includes use on level 1 as an area term (list O). For term set options see list B.

Pennsylvanian
Includes use on level 1 as an age term (list E); on level 2 under paleoterms, e.g. paleoecology, paleogeography, paleomagnetism.

—— CR see also Carboniferous; see also under geochronology under ——; see also under stratigraphy under ——

Pentamerida
Includes use on level 2 under Brachiopoda(1). See list F.

Pentoxylales
Term introduced in 1981. Includes use on level 2 under gymnosperms(1). See list F.

peridotites
Term introduced in 1981. Includes use on level 2 under igneous rocks(1). See list H.

—— CR see under igneous rocks

periglacial features
Includes use on level 2 under glacial geology(1).

—— CR see under geomorphology; glacial geology

periodicity
Includes use on level 2 under orogeny(1); on level 3 under earthquakes(1) and under volcanology(1).

Perisphinctida
Term introduced in 1981. Includes use on level 2 under Mollusca(1). See list F.

Perissodactyla
Includes use on level 2 under Mammalia(1). See list F.

permafrost
A level 1 term as of 1978. Used for geotechnical studies of permafrost; geomorphological studies should be under geomorphology. Term set options are:
  active layer
    subtopic
  classification
    subtopic
  creep
    subtopic
  engineering properties
    subtopic
  experimental studies
    subtopic
  frost action
    subtopic
  frost heaving
    subtopic
  site exploration
    subtopic
  solifluction
    subtopic
  theoretical studies
    subtopic

—— CR see also under engineering geology under ——

Permian
Includes use on level 1 as an age term (list E); on level 2 under paleoterms, e.g. paleoecology, paleogeography, paleomagnetism. Above Carboniferous, below Triassic (Mesozoic).

—— CR see also under geochronology under ——; see also under stratigraphy under ——

Persian Gulf
Between Arabian Peninsula on the W and S, and Iran on E. Includes use on level 1 as an area term (list O). For term set options see list B.

Peru
Includes use on level 1 as an area term (list O). For term set options see list B.

petrography
Includes use on level 2 under sedimentary rocks(1) for microscopic studies only; on level 2 under metamorphic rocks(1) and under igneous rocks(1).

petroleum
Includes use on level 1 and 2 as a commodity term (list C).

—— CR see also under economic geology under ——

petroleum engineering
Term introduced in 1978 on level 2 under engineering geology(1).

petrology
Treated as a whole. For studies on igneous or metamorphic rocks. Includes use on level 1 (list A); on level 2 under area terms(1), bibliography(1), continental shelf(1), education(1), intrusions(1), lava(1), and sediments(1). See list B. If level 1, term set options are:
  topic [applications, bibliography, catalogs, classification, concepts, education, experimental studies, general, history, instruments, methods, nomenclature, objectives, philosophy, practice, principles, symposia, textbooks, theoretical studies]
    subtopic

—— see sedimentary petrology; structural petrology

Phacopida
Includes use on level 2 under Trilobita(1). See list F.

Phaeodarina
Includes use on level 2 under Radiolaria(1). See list F.

Phaeophyta
Including brown algae and seaweed. Includes use on level 2 under algae(1). See list F.

Phanerozoic
Includes use on level 1 and 2 as an age term (list E).

—— CR see also Cambrian; Carboniferous; Cenozoic; Cretaceous; Devonian; Eocene; Holocene; Jurassic; Mesozoic; Miocene; Mississippian; Neogene; Oligocene; Ordovician; Paleocene; Paleogene; Paleozoic; Pennsylvanian; Permian; Pleistocene; Pliocene; Quaternary; Silurian; Tertiary; Triassic; see also under geochronology under ——; see also under stratigraphy under ——

phase equilibria
For laboratory studies only. Includes use on level 1 (list A) and on level 2 under crystal chemistry(1) and crystal growth(1). If 1, term set options are:
  material [e.g. igneous rocks, silicates; only 1st and 2nd level terms]
  (system in chemical symbols) or topic
  topic [anomalies, concepts, experimental studies, interpretation, melting, theoretical studies]

(system) or subtopic

phenols
Includes use on level 2 under organic materials(1).

phenomena *see* electrical phenomena; surface phenomena

Philippine Islands
Includes use on level 1 as an area term (list O). For term set options see list B.

Philippine Sea
Includes use on level 1 as an area term (list O). That part of the W Pacific Ocean with the Philippines Islands on the W, Taiwan and the Ryukyus on the NW, and the U.S. Trust Territory of the Pacific Islands on the E and SE.

philosophy
Use term under discipline. Includes use as a level 2 or 3 term appropriate to a large number of topics, e.g. on level 2 under geology(1). See list G.

Pholadomyida
Term introduced in 1981. Includes use on level 2 under Mollusca(1). See list F.

Pholidota
Includes use on level 2 under Mammalia(1). See list F.

phonolites
Term introduced in 1981. Includes use on level 2 under igneous rocks(1). See list H.

—— *CR see under* igneous rocks

Phoronida
Term introduced in 1981. Includes use on level 2 under worms(1). See list F.

phosphate composition
Term introduced in 1978. Included use on level 2 under nodules(1) until 1981.

phosphate deposits
Term introduced in 1981. Includes use on level 1 and 2 as a commodity term (list C).

—— *CR see also under* economic geology *under* ——

phosphates
Includes use on level 2 under minerals(1). See list L.

—— *CR see under* minerals

phosphides
As of 1981, includes use on level 2 under minerals(1). See list L. Before 1981, included use on level 3 under minerals(1) native elements and alloys(2).

—— *CR see under* alloys *under* minerals

phosphorus
Includes use on level 1 and 2 as a chemical element (list D); on level 3 under soils(1).

photogeologic methods
Includes use on level 2 under mineral exploration(1).

photography
Includes use on level 2 under micropaleontology(1).

—— *see* aerial photography

phyllites
Includes use on level 2 under metamorphic rocks(1). See list J.

Phylloceratida
Term introduced in 1981. Includes use on level 2 under Mollusca(1). See list F.

phyllonites
Includes use on level 2 under metamorphic rocks(1).

piles
As of 1978, term is used on level 2 under foundations(1).

—— *CR see under* foundations

pingos *CR see under* periglacial features *under* geomorphology; glacial geology

Pinnipedia
Term introduced in 1981. Includes use on level 2 under Mammalia(1). See list F.

pipelines
As of 1978, term is used on level 2 under marine installations(1).

pipes
Includes use on level 2 under intrusions(1).

Pisces
Includes use on level 1 and 2 as a fossil term (list F).

placers
Includes use on level 1 (list A). Used for a surficial mineral deposit formed by mechanical concentration of mineral particles from weathered debris. See also list C for type of commodity. Term set options are:
kind of placer [e.g. diamonds, gold ores, heavy mineral deposits, platinum ores, tin ores; first level terms only]
area
topic [detection, distribution, exploration, genesis, identification, patterns, sampling]
subtopic

—— *CR see also* heavy minerals

Placodermi
Includes use on level 2 under Pisces(1). See list F.

Placodontia
Term introduced in 1981. Includes use on level 2 under Reptilia(1). See list F.

plagioclase
As of 1981, includes use in combination with framework silicates (i.e. framework silicates, plagioclase) on level 2 under minerals(1). See list L. Before 1981, included use on level 3 under minerals(1) framework silicates, feldspar group(2).

planar bedding structures
Includes use on level 2 under sedimentary structures(1). See list K.

planetary interiors
Term introduced in 1981. Includes use on level 2 under individual planets.

planetology
For studies of more than one planet, the relationship between planets, or the solar system as a whole. See also names of planets. Includes use on level 1 (list A). Term set options are:
topic [atmosphere, bibliography, concepts, cosmic dust, cosmic rays, cosmochemistry, methods, nomenclature, principles, techniques, theoretical studies]
subtopic

—— *CR see also* Jupiter; Mars; Mercury Planet; Moon; Saturn; Venus

planning
As of 1978 term is used on level 2 under highways(1) and land use(1).

—— *see* regional planning; urban planning

Plantae
Includes use on level 1 and 2 as a fossil term (list F). In indexing, used in a general sense when specific plants are unknown.

—— *CR see also* algae; angiosperms; bacteria; bryophytes; fungi; gymnosperms; ichnofossils; lichens; palynomorphs; problematic fossils; Protista; pteridophytes; thallophytes

plasma instabilities
Includes use on level 2 under magnetosphere(1).

plasma motion
Includes use on level 2 under magnetosphere(1).

plasmapause
Includes use on level 2 under magnetosphere(1).

plasticity
As of 1978, term is used on level 2 under rock mechanics(1).

—— *CR see under* —— *under* deformation; *see under* rock mechanics; soil mechanics

plate geometry
Term introduced in 1981. Includes use on level 2 under plate tectonics(1).

plate tectonics
Includes use on level 1, as of 1976. Used for global tectonics based on an Earth model characterized by a small number of large, broad, thick plates. Before 1976, articles discussing plate tectonics were included under tectonophysics(1). See list B. Term set options are:
age
age term (List E)

topic [concepts, effects, evolution, indicators, island arcs, mechanism, movement, obduction, plate geometry, processes, rifting, subduction]
    subtopic
—— *CR see also under* tectonophysics *under* ___

platinum
Includes use on level 1 and 2 as a chemical element (list D). As of 1981, use platinum ores for platinum as a commodity. Before 1981, included use on level 1 as a commodity term.

platinum ores
Term introduced in 1981. Includes use on level 1 and 2 as a commodity term (list C).
—— *CR see also under* economic geology *under* ___

Platycopida
Term introduced in 1981. Includes use on level 2 under Ostracoda(1). See list F.

Platyrrhina
Term introduced in 1981. Includes use on level 2 under Mammalia(1). See list F.

Pleistocene
Glacial epoch. World. Above Pliocene (Tertiary), below Holocene. Includes use on level 1 as an age term (list E); on level 2 under paleoterms, e.g. paleoecology, paleogeography; on level 3 under age terms(1).
—— *CR see also under* geochronology *under* ___; *see also under* stratigraphy *under* ___

Plesiosauria
As of 1981, includes use on level 2 under Reptilia(1). See list F. Before 1981, included use on level 3 under Reptilia(1) Euryapsida(2).

Pliocene
World. Above Miocene, below Pleistocene (Quaternary). Includes use on level 1 as an age term (list E); on level 2 under paleo- terms, e.g. paleoecology, paleogeography; on level 3 under age terms(1).
—— *CR see also under* geochronology *under* ___; *see also under* stratigraphy *under* ___

plugs
Includes use on level 2 under intrusions(1).

Pluto
Includes use on level 1 and 2. See entry under Moon for term set options.

plutonic rocks
Term introduced in 1978. Includes use on level 2 and under igneous rocks(1). See list H.

plutonium
Includes use on level 1 and 2 as a chemical element (list D).

plutons
Includes use on level 2 under intrusions(1).
—— *CR see under* intrusions

Pogonophora
Term introduced in 1981. Includes use on level 2 under worms(1). See list F.

Poland
Includes use on level 1 as an area term (list O). For term set options see list B.

polar wandering
As of 1978, term is used on level 2 under paleomagnetism(1).

polarography *CR see under* methods *under* chemical analysis

pole positions
Includes use on level 2 under paleomagnetism(1).

pollutants
As of 1978, term is used on level 2 under pollution(1).

pollution
A level 1 term as of 1978. Used for geological studies on pollution of the environment. Includes use on level 2 under impact statements(1) and nuclear facilities(1). If 1, term set options are:
    air
        subtopic
    case studies
        subtopic
    causes
        subtopic
    concepts
        subtopic
    controls
        subtopic
    detection
        subtopic
    effects
        subtopic
    experimental studies
        subtopic
    field studies
        subtopic
    ground water
        subtopic
    human ecology
        subtopic
    impact statements
        subtopic
    metals
        subtopic
    oil spills
        subtopic
    pollutants
        subtopic
    surface water
        subtopic
    waste disposal

    subtopic
    water
        subtopic
—— *CR see also* reclamation; waste disposal; *see also under* environmental geology *under* ___
—— *see* thermal pollution

polonium
Includes use on level 1 and 2 as a chemical element (list D).

Polychaetia
As of 1981, includes use on level 2 under worms(1). See list F. Before 1981, included use on level 2 under Annelida(1).

polymetallic ores
Includes use on level 1 and 2 as a commodity term (list C).
—— *CR see also* gold ores; silver ores; *see also under* economic geology *under* ___

polymetamorphism
Includes use on level 2 under metamorphism(1).

Polynesia
Collective name for islands of the central and SE Pacific Ocean. Index island and island groups as applicable. Includes use on level 1 or 2 as an area term (list O). If 1, see term set options under list B.
—— *CR see also* Samoa; Society Islands; Tahiti; Tonga

Polyplacophora
Includes use on level 2 under Mollusca(1). See list F.

Polyprotodontia
Term introduced in 1981. Includes use on level 2 under Mammalia(1). See list F.

Pongidae
Family. As of 1981, includes use on level 2 under Mammalia(1). See list F. Before 1981, included use on level 3 under Mammalia(1) Primates(2).

pore water
Includes use on level 2 under sediments(1).

Porifera
Includes use on level 1 and 2 as a fossil term (list F).

Portugal
Includes use on level 1 as an area term (list O). For term set options see list B.

possibilities
Includes use on level 2 and 3. As of 1981, restricted to economic geology, especially referring to the economic potential of a mineral deposit. Before 1981, included use as a general term.

potash
Includes use on level 1 as a commodity term (list C). As of 1981, includes use on level 3 under sedimentary rocks(1) chemically precipitated rocks(2). See list I.

—— *CR see also under* economic geology *under* ——

potassium
Includes use on level 1 and 2 as a chemical element (list D).

practice
To be used for geology as a profession after 1975. Includes use as a level 2 or 3 term appropriate to a large number of topics, e.g. on level 2 under geochemistry(1), geology(1), and economic geology(1). See list G.

Praecardiida
Term introduced in 1981. Includes use on level 2 under Mollusca(1). See list F.

praseodymium
Includes use on level 1 and 2 as a chemical element (list D.)

pre-Neanderthal
Term introduced in 1981. Includes use on level 2 under Mammalia(1). See list F.

Precambrian
Includes use on level 1 as an age term (list E); on level 2 under paleoterms, e.g. paleoecology, paleogeography, paleomagnetism.
—— *CR see also* Archean; Proterozoic; *see also under* geochronology *under* ——; *see also under* stratigraphy *under* ——

precipitated   *see*   chemically precipitated rocks

precipitation
As of 1978, term refers to geochemistry. For meteorology, use atmospheric precipitation. Includes use on level 2 under sedimentation(1).
—— *see* atmospheric precipitation; particle precipitation

prediction
Includes use on level 2 under earthquakes(1).

preferred orientation
Includes use on level 2 under structural analysis(1).

preparation
Includes use as a level 2 or 3 term appropriate to a large number of topics, e.g. on level 2 under coal(1). See list G.

preservation
As of 1978, term is used on level 2 under land use(1).

pressure *see* earth pressure

Priapulida
Term introduced in 1981. Includes use on level 2 under worms(1). See list F.

primary structures
Includes use on level 2 under sedimentary structures(1). See list K.

Primates
Includes use on level 2 under Mammalia(1). See list F.

Prince Edward Island
Island in the Gulf of Saint Lawrence constituting a province. Includes use on level 1 as an area term (list O). For term set options see list B.

principles
Includes use as a level 2 or 3 term appropriate to a large number of topics, e.g. on level 2 under geology(1). See list G.

probe *see* electron probe

problematic fossils
Term introduced in 1978. Includes use on level 1 and 2.

problematic microfossils
Term introduced in 1981. Includes use on level 2 under problematic fossils(1). See list F.

Proboscidea
Includes use on level 2 under Mammalia(1). See list F.

processes
Includes use as a level 2 or 3 term appropriate to a large number of topics, e.g. on level 2 under geochemistry(1), sedimentation(1), and under mineral deposits, genesis(1). See list G.
—— *see* cyclic processes

production
Includes use on level 2 and 3 under commodity terms, e.g. on level 2 under energy sources(1). See list C.

production control
Includes use on level 2 under mining geology(1).

Proganosauria
Term introduced in 1981. Includes use on level 2 under Reptilia(1). See list F.

prograde metamorphism
Term introduced in 1978. Includes use on level 2 under metamorphism(1).

programs
Includes use on level 2 under mineral exploration(1); on level 3 under automatic data processing(1) and education(1).

Prolecanitida
Term introduced in 1981. Includes use on level 2 under Mollusca(1). See list F.

promethium
Includes use on level 1 and 2 as a chemical element (list D).

propagation
Includes use on level 3 under seismology(1) elastic waves(2). As of 1981, includes use on level 2 under ionosphere(1) and magnetosphere(1).

properties
Includes use in combination with materials (i.e. materials, properties) on level 2 under engineering geology(1); as a level 2 or 3 term appropriate to a large number of topics; for physical or chemical properties on level 2 or 3 under commodity terms(1). See list C (commodities) and list G (general terms).
—— *see* engineering properties; optical properties; surface properties

propylite
As of 1978, term is used on level 2 under metasomatic rocks(1). Before 1978, included use on level 3 under igneous rocks(1) andesite-rhyolite family(2).

Prosimii
Term introduced in 1981. Includes use on level 2 under Mammalia(1). See list F.

prospecting *see* geochemical prospecting; mineral prospecting

protactinium
Includes use on level 1 and 2 as a chemical element (list D).

Proterozoic
As of 1978, includes use on level 1 and 2 (list E).
—— *CR see also* Precambrian; *see also under* geochronology *under* ——; *see also under* stratigraphy *under* ——

Proteutheria
Term introduced in 1981. Includes use on level 2 under Mammalia(1). See list F.

Protista
Includes use on level 1 and 2 as a fossil term (list F).

Protozoa *CR see* Protista

provenance
Includes use on level 2 under sedimentation(1) and under sediments(1).

Psiloceratida
Term introduced in 1981. Includes use on level 2 under Mollusca(1). See list F.

Psilopsida
Including Psilophytales. Includes use on level 2 under pteridophytes(1). See list F.

Psocopteroida
Term introduced in 1981. Includes use on level 2 under Insecta(1). See list F.

Pteridophyllen
Includes use on level 2 under pteridophytes(1). See list F.

pteridophytes
Includes use on level 1 and 2 as a fossil term (list F).

Pteriina
Term introduced in 1981. Includes use on level 2 under Mollusca(1). See list F.

Pterobranchia
Includes use on level 1 and 2 as a fossil term (list F).

Pteropoda
As of 1981, includes use on level 2 under Mollusca(1). See list F. Before 1981, included use on level 3 under Mollusca(1) Gastropoda(2).

Pterosauria
As of 1981, includes use on level 2 under Reptilia(1). See list F. Before 1981, included use on level 3 under Reptilia(1) Archosauria(2).

Ptychopariida
Includes use on level 2 under Trilobita(1). See list F.

Puerto Rico
Includes use on level 1 as an area term (list O). For term set options see list B.

pumice CR see also under ___ under igneous rocks

pumice deposits
Term introduced in 1981. Includes use on level 1 and 2 as a commodity term (list C).
—— CR see also under economic geology under ___

Pyrenees
Mountain range extending from the Bay of Biscay to the SW coast of the Gulf of Lion. Index countries as applicable. Includes use on level 1 as an area term (list O). For term set options see list B.

pyrite ores
Term introduced in 1981. Includes use on level 1 and 2 as a commodity term (list C).
—— CR see also under economic geology under ___

pyroclastics
Includes use on level 3 under sedimentary rocks(1) clastic rocks(2), under sediments(1) clastic sediments(2), and as of 1981, on level 2 under igneous rocks(1). See lists H, I, and N. Before 1981, included use on level 3 under igneous rocks(1) pyroclastics and glasses(2).
—— CR see under igneous rocks

Pyrotheria
Term introduced in 1981. Includes use on level 2 under Mammalia(1). See list F.

pyroxene see alkalic pyroxene

pyroxene group
Includes use in combination with chain silicates (i.e. chain silicates,pyroxene group) on level 2 under minerals(1). See list L.

Pyrrhophyta
As of 1981, includes use on level 2 under algae(1). See list F. Before 1981, included use on level 3.

Qatar
Includes use on level 1 as an area term (list O). For term set options see list B.

quartz CR see under ___ under ___

quartz crystal
Includes use on level 1 as a commodity term (list C).
—— CR see also under economic geology under ___

quartz diorites
Term introduced in 1981. Includes use on level 2 under igneous rocks(1). See list H.
—— CR see under igneous rocks

quartzites
Term introduced in 1981. Includes use on level 2 under metamorphic rocks(1). See list J.

Quaternary
Includes use on level 1 as an age term (list E); on level 2 under paleoterms, e.g. paleoecology, paleogeography, paleomagnetism. Consists of Pleistocene and Holocene.
—— CR see also Holocene; Pleistocene; see also under geochronology under ___; see also under stratigraphy under ___

Quebec
Includes use on level 1 as an area term (list O). For term set options see list B.

Queensland
Includes use on level 1 as an area term (list O). For term set options see list B.

racemization
Includes use on level 2 under geochronology(1).

radiation see electromagnetic radiation; particle radiation

radiation damage
Includes use on level 2 under geochronology(1).
—— CR see under geochronology

radioactive dating CR see absolute age

radioactive waste
As of 1978, term is used on level 2 under waste disposal(1).

radioactivity
For natural or induced radioactivity. Includes use on level 2 under well-logging(1) and on level 3 under soils(1), heat flow(1), and pollution(1).

radioactivity methods
Includes use on level 2 under geophysical methods(1).

radioactivity surveys CR see under geophysical surveys under ___

Radiolaria
Includes use on level 1 and 2 as a fossil term (list F).

radiolarians
Term introduced in 1981 as the common name for Radiolaria. See list F. To be used for Radiolaria when they are used as tools in stratigraphy. Includes use on level 1. If 1, term set options are:
    biostratigraphy
    age [List E]

radium
Includes use on level 1 and 2 as a chemical element (list D).

radon
Includes use on level 1 and 2 as a chemical element (list D).

rare earth deposits
Term introduced in 1981. Includes use on level 1 and 2 as a commodity term (list C).
—— CR see also under economic geology under ___

rare earths
Includes use on level 1 and 2 as a chemical element (list D). As of 1981, use rare earth deposits for rare earths as a commodity. Before 1981, included use on level 1 as a commodity term.
—— CR see also cerium; dysprosium; erbium; europium; gadolinium; holmium; lanthanum; lutetium; neodymium; praesodymium; prometheum; samarium; scandium; terbium; thulium; ytterbium; yttrium

rates
Includes use as a level 2 or 3 term appropriate to a large number of topics. See list G. Before 1982, included use on level 2 under sedimentation(1). As of 1982, use sedimentation rates under sedimentation(1).

ratios
Includes use on level 2 under isotopes(1); on level 3 under name of element(1) isotopes(2) and under absolute age(1). See list D.

rays see cosmic rays

recharge
Includes use on level 2 under ground water(1).

reclamation
A level 1 term as of 1978. Used for geological studies on the reclamation of the natural environment. Includes use on level 2 under impact statements(1) and land use(1). If 1, term set options are:
    environment
    beaches
    drainage basins
    mines
    open-pit mining

strip mining
natural resources
type (floods, land, soils, waste water, etc.)
topic (experimental studies, methods, practice, programs)
subtopic
—— *CR see also* conservation; land use; pollution; *see also under* environmental geology *under* ——

recreation
As of 1978, term is used on level 2 under land use(1).

Red Sea
Between NE Africa and the Arabian Peninsula connecting the Mediterranean Sea with the Indian Ocean. Includes use on level 1 or 2 as an area term (list O). If 1, see term set options under list B.

Red Sea region
Introduced as a level 1 and 2 area term (list O) in 1976. This is an artificial term used to indicate the coastal region immediately adjacent to the Red Sea and the immediate littoral zone. For term set options see list B.

Redlichiida
Includes use on level 2 under Trilobita(1). See list F.

reefs
Includes use on level 1 (list A); on level 3 for type of environment under sedimentation(1). If used on level 1, term set options are:
topic [age, distribution, ecology, evolution, lithofacies, paleoecology]
subtopic (no area term)
—— *CR see also under* oceanography *under* ——; *see also under* sedimentary petrology *under* ——

refinement
As of 1978, term is used on level 2 under crystal structure(1).

regimes *see* water regimes

regional geology *CR see* areal geology under the appropriate area term

regional metamorphism
Term introduced in 1978. Includes use on level 2 under metamorphism(1).

regional patterns
Includes use on level 2 under heat flow(1).

regional planning
As of 1978, term is used on level 2 under land use(1).

remote sensing
A level 1 term as of 1978. Used for both methods and applications. Includes use on level 2 under mineral exploration(1) and land use(1), and on level 3 in area sets under geophysical surveys(2). If 1, term set options are:

aerial photography
subtopic
applications
subtopic
automatic data processing
subtopic
imagery
subtopic
instruments
subtopic
interpretation
subtopic
methods
subtopic
photogeologic methods
subtopic
—— *CR see also* geophysical methods; geophysical surveys; *see also under* geophysical surveys *under* ——

report *see* annual report

reptiles
Term introduced in 1981 as the common name for Reptilia. See list F. To be used for Reptilia when they are used as tools in stratigraphy. Includes use on level 1. If 1, term set options are:
biostratigraphy
age [List E]

Reptilia
Includes use on level 1 and 2 as a fossil term (list F).

research
Includes use on level 2 under geology(1).
—— *see* current research

reserves
Includes use on level 2 and 3 under commodity terms(1), e.g. on level 2 under coal(1). See list C.

reservoirs
A level 1 term as of 1978. Before 1978, included use on level 2 under engineering geology(1). Used for geological studies on surface reservoirs only. For subsurface reservoirs use petroleum engineering under engineering geology.
construction
subtopic
dams
subtopic
design
subtopic
earthquakes
subtopic
experimental studies
subtopic
feasibility studies
subtopic
field studies
subtopic
maintenance
subtopic
seepage

subtopic
site exploration
subtopic
storage
subtopic
—— *CR see also under* engineering geology *under* ——

residues *see* organic residues

resources
Includes use on level 2 and 3 under commodity terms(1), e.g. on level 2 under energy sources(1). See list C.
—— *see* fuel resources; mineral resources; natural resources; water resources

retrograde metamorphism
Term introduced in 1978. Includes use on level 2 under metamorphism(1).

Reunion
One of the Mascarene Islands and a French overseas territory 425 miles E of the Malagasy Republic. Includes use on level 1 as an area term (list O). For term set options see list B.

reversals
Includes use on level 2 under paleomagnetism(1).

reviews *see* book reviews

Rhapdopleurida
Includes use on level 2 under Pterobranchia(1). See list F.

rhenium
Includes use on level 1 and 2 as a chemical element (list D).

rheomorphism
Includes use on level 2 under metamorphism(1).

Rhipidistia
As of 1981, includes use on level 2 under Pisces(1). See list F. Before 1981, included use on level 3 under Pisces(1) Osteichthyes(2).

Rhode Island
Includes use on level 1 as an area term (list O). For term set options see list B.

Rhodesia *CR see* Zimbabwe

rhodium
Includes use on level 1 and 2 as a chemical element (list D).

Rhodophyta
Including red algae. Includes use on level 2 under algae(1). See list F.

Rhynchocephalia
Term introduced in 1981. Includes use on level 2 under Reptilia(1). See list F.

Rhynchonellida
Includes use on level 2 under Brachiopoda(1). See list F.

rhyodacites
Term introduced in 1981. Includes use on level 2 under igneous rocks(1). See list H.
—— *CR see under* igneous rocks

rhyolite *CR see under* _____ *under* igneous rocks

rhyolites
Term introduced in 1981. Includes use on level 2 under igneous rocks(1). See list H.

_____ *CR see under* igneous rocks

ridges *see* mid-ocean ridges

rift zones *CR see under* _____ *under* faults; plate tectonics; *see under* ocean floors

rifting
Includes use on level 2 under plate tectonics(1).

ring complexes
As of 1978, term is used on level 2 under intrusions(1).

ring silicates
Includes use on level 2 under minerals(1). See list L.

_____ *CR see under* minerals

rings *see* tree rings

ripple marks *CR see under* _____ *under* sedimentary structures

rivers and streams
Includes use on level 2 under hydrology(1) and under waterways(1).

Robertinacea
Superfamily. Includes use on level 2 under foraminifera(1).

rock bursts
As of 1978, term is used on level 2 under geologic hazards(1).

rock mechanics
A level 1 term as of 1978. Includes use on level 2 under nuclear facilities(1), tunnels(1), and underground installations(1). Used for geotechnical studies. If 1, term set options are:
applications
 subtopic
case studies
 subtopic
concepts
 subtopic
deformation
 subtopic
elasticity
 subtopic
excavations
 subtopic
experimental studies
 subtopic
field studies
 subtopic
frost action
 subtopic
failures
 subtopic
materials, properties
 subtopic
methods
 subtopic
plasticity
 subtopic
site exploration

subtopic
techniques
 subtopic
theoretical studies
 subtopic

_____ *CR see also* foundations; soil mechanics; underground installations

rockfalls
As of 1978, term is used on level 2 under slope stability(1).

rocks *see* carbonate rocks; chemically precipitated rocks; clastic rocks; hypabyssal rocks; igneous rocks; metaigneous rocks; metamorphic rocks; metaplutonic rocks; metasedimentary rocks; metasomatic rocks; metavolcanic rocks; plutonic rocks; sedimentary rocks; volcanic rocks

Rocky Mountains
Mountain system in W North America extending from N Alaska to the Mexican frontier. Index Alaska, and countries as applicable. Includes use on level 1 as an area term (list O). For term set options see list B.

Rodentia
Includes use on level 2 under Mammalia(1). See list F.

Romania
Includes use on level 1 as an area term (list O). For term set options see list B.

Rosidae
Term introduced in 1981. Includes use on level 2 under angiosperms(1). See list F.

Rostroconchia
Includes use on level 2 under Mollusca(1). See list F.

Rotaliacea
Includes use on level 2 under foraminifera(1). See list F.

Rotaliina
Includes use on level 2 under foraminifera(1). See list F.

rubidium
Includes use on level 1 and 2 as a chemical element (list D).

Rudistae
As of 1981, includes use on level 2 under Mollusca(1). See list F. Before 1981, included use on level 3 under Mollusca(1) Bivalvia(2).

Rugosa
Includes use on level 2 under Coelenterata(1). See list F.

Ruminantia
As of 1981, includes use on level 2 under Mammalia(1). See list F. Before 1981, included use on level 3 under Mammalia(1) Artiodactyla(2).

Russia *CR see* USSR

ruthenium
Includes use on level 1 and 2 as a chemical element (list D).

Rwanda
Formerly Ruanda which was part of the Belgium trust territory of Ruanda-Urundi. Includes use on level 1 as an area term (list O). For term set options see list B.

Sacoglossa
Term introduced in 1981. Includes use on level 2 under Mollusca(1). See list F.

Sahara
Vast arid region extending across North Africa from the Atlantic Ocean to the Red Sea. Index countries as applicable. Includes use on level 1 as an area term (list O). For term set options see list B. From 1977-80, documents on the Spanish Sahara were indexed under Sahara as the level-1 term. As of 1981, use Western Sahara for the Spanish Sahara.

Saint Pierre and Miquelon
Term introduced in 1981. Consists of two islands in the Atlantic Ocean just S of Newfoundland. Includes use on level 1 as an area term (list O). For term set options see list B.

salinity
Includes use on level 2 under soils(1) and under paleoecology(1).

salt
Includes use on level 1 and 2 as a commodity term (list C); on level 2 under mineral deposits, genesis(1). As of 1981, includes use on level 3 under sedimentary rocks(1) chemically precipitated rocks(2). See list I.

_____ *CR see also under* _____ *under* sedimentary rocks; sediments; *see also under* economic geology *under* _____

salt domes *CR see under* _____ *under* salt tectonics

salt tectonics
Includes use on level 1 (list A). Used for the study of the structure and mechanism of emplacement of salt domes. Term set options are:
 topic [causes, evolution, interpretation, mechanism, processes]
 subtopic (no area term)

_____ *CR see also under* structural geology *under* _____

salt-water intrusion
Includes use on level 2 under ground water(1).

_____ *CR see under* ground water

samarium
Includes use on level 1 and 2 as a chemical element (list D).

Samoa
Group of volcanic islands in SW central Pacific Ocean. American Samoa is in E part of group, and independ-

ent Western Samoa comprises the W part. Includes use on level 1 as an area term (list O). For term set options see list B.

sampling
Includes use on level 2 under placers(1).

San Marino
Term introduced in 1981 as a level 1 area term (list O). Republic in the Apennines, N part of Italian peninsula near the Adriatic coast.

sands
As of 1981, includes use on level 1 and 2 as a commodity term (list C) for glass, ceramic, chemical use, etc. For sand used as a construction material, use gravel deposits. Before 1981, included use as a type of material in engineering geology; as of 1981, use sand.

—— *CR see also under* economic geology *under* ——

—— *see* oil sands; tar sands

sandstone *CR see also under* —— *under* sedimentary rocks

sandstone deposits
Term introduced in 1981. Includes use on level 1 and 2 as a commodity term (list C).

—— *CR see also under* economic geology *under* ——

Sardinia
Island and autonomous region in the Mediterranean Sea. Includes use on level 1 as an area term (list O). For term set options see list B.

Saskatchewan
Includes use on level 1 as an area term (list O). For term set options see list B.

satellite measurements
Includes use on level 2 under geodesy(1).

satellites
Used for natural satellites other than the Earth's Moon. As of 1981, includes use on level 2 under specific planets, e.g. Jupiter(1). Before 1981, included use on level 3 in planet sets. For artificial satellites, use remote sensing.

Saturn
Includes use on level 1 and 2. See entry under Moon(1) for term set options.

Saudi Arabia
Includes use on level 1 as an area term (list O). For term set options see list B.

Saurischia
As of 1981, includes use on level 2 under Reptilia(1). See list F. Before 1981, included use on level 3 under Reptilia(1) Archosauria(2).

Sauropterygia
Term introduced in 1981. Includes use on level 2 under Reptilia(1). See list F.

scales *see* time scales

Scandentia
Term introduced in 1981. Includes use on level 2 under Mammalia(1). See list F.

Scandinavia
Region comprising Denmark, Norway, Sweden, and Finland. Iceland sometimes included. Index countries as applicable. Includes use on level 1 as an area term (list O). For term set options see list B.

scandium
Includes use on level 1 and 2 as a chemical element (list D).

Scaphopoda
Includes use on level 2 under Mollusca(1). See list F.

scapolite group
Includes use in combination with framework silicates (i.e. framework silicates, scapolite group) on level 2 under minerals(1). See list L.

scattering *see* absorption and scattering

schist *CR see under* —— *under* metamorphic rocks

schistosity *CR see under* —— *under* foliation; structural analysis

schists
Includes use on level 2 under metamorphic rocks(1). See list J.

scintillations
Includes use on level 2 under ionosphere(1).

Sciuromorpha
Term introduced in 1981. Includes use on level 2 under Mammalia(1). See list F.

Scleractinia
Includes use on level 2 under Coelenterata(1). See list F.

scolecodonts
Includes use on level 2 under worms(1). See list F.

Scotland
Includes use on level 1 as an area term (list O). For term set options see list B.

Scyphozoa
Includes use on level 2 under Coelenterata(1). See list F.

sea ice
Includes use on level 2 under oceanography(1).

Sea of Japan *CR see* Japan Sea
Sea of Okhotsk *CR see* Okhotsk Sea

sea water
Includes use on level 1 (list A); on level 2 under isotopes(1). Used for studies on the composition and properties of the water of the

oceans. For ancient water, see sets such as sedimentation or geochemistry. Term set options are:
  composition
    subtopic
  evolution
    subtopic
  experimental studies
    subtopic
  genesis
    subtopic
  geochemistry
    subtopic
  theoretical studies
    subtopic

sea-floor spreading
Includes use on level 1 (list A); on level 2 under tectonophysics(1). Used for topics related to the hypothesis that the oceanic crust is increasing by convective upwelling of magma along the mid-ocean ridges or world rift system. If 1, term set options are:
  topic [causes, concepts, evolution, mechanism, rates]
  subtopic (no area term)

—— *CR see also* continental drift; plate tectonics; *see also under* tectonophysics *under* ——

sea-level fluctuations *CR see* changes of level

seamounts
Includes use on level 2 under ocean floors(1).

—— *CR see under* ocean floors

seawalls
Term introduced in 1978 on level 2 under shorelines(1).

secondary structures
Includes use on level 2 under sedimentary structures(1). See list K.

sedimentary petrology
Treated as a whole. Includes use on level 1 (list A); on level 2 under area terms(1), bibliography(1), education(1), and symposia(1). If level 1, term set options are:
  topic [applications, bibliography, catalogs, classification, concepts, education, history, instruments, methods, nomenclature, philosophy, practice, principles, symposia, techniques, textbooks]
  subtopic

sedimentary rocks
See list I. Includes use on level 1 (list A); on level 2 under paragenesis(1) and weathering(1). If level 1, term set options are:
  rock group [carbonate rocks, chemically precipitated rocks, clastic rocks, organic residues]
  rock name
  topic [classification, composition, diagenesis, environmental analysis, geochemistry, lithofacies, lithos-

tratigraphy, petrography, petrology, pore water, properties, provenance, textures]
subtopic [e.g. fabric, grains, grain size, rounding, sorting, surface textures, etc.]
—— CR see also sedimentary structures; sedimentation; sediments

sedimentary structures
See list K for types and names of structures. Includes use on level 1 (list A). Used for a structure in a sedimentary rock, formed either contemporaneously with deposition (primary) or subsequently to deposition (secondary). Term set options are:
topic [classification, environmental analysis, genesis, interpretation, nomenclature, patterns]
subtopic
type of structure [bedding plane irregularities, biogenic structures, cylindrical structures, planar bedding structures, primary structures, secondary structures, soft sediment deformation, turbidity current structures]
specific type (List K)
—— CR see also sedimentary rocks; sediments

sedimentation
Includes use on level 1 (list A); on level 2 under paleoecology(1). Used for the act or process of forming or accumulating sediment in layers, including all processes from transport through diagenesis. If level 1, term set options are:
environment
type of environment [coastal environment, deltaic environment, eolian environment, estuarine environment, intertidal environment, lacustrine environment, lagoonal environment, marine environment, paludal environment, reefs, terrestrial environment]
topic [controls, cyclic processes, deposition, diagenesis, precipitation, processes, provenance, sedimentation rates]
subtopic
transport
type of transport [glacial transport, ice-rafting, marine transport, stream transport, turbidity currents, wind transport]

sedimentation rates
Term introduced in 1982. Includes use on level 2 under sedimentation(1).

sediments
See list N. Includes use on level 1 (list A); on level 2 under weathering(1). Used for unconsolidated solid fragmental material, or a mass

of such material, that originates from weathering of rocks. If 1, term set options are:
sediment group [carbonate sediments, clastic sediments, marine sediments, organic residues]
sediment name
topic [classification, composition, diagenesis, distribution, environmental analysis, genesis, geochemistry, lithofacies, lithostratigraphy, petrography, petrology, pore water, properties, provenance, textures]
subtopic [e.g. grain size]
—— CR see also sedimentary rocks; sedimentary structures; sedimentation
—— see carbonate sediments; clastic sediments; marine sediments

seepage
Includes use on level 2 under engineering geology(1).
—— CR see under ____

seismic intensity
Term introduced in 1981. Includes use on level 2 under earthquakes(1).

seismic methods
Includes use on level 2 under geophysical methods(1).

seismic sources
Includes use on level 2 under seismology(1).

seismic surveys
Includes use on level 3 under area terms(1) geophysical surveys(2).
—— CR see under geophysical surveys under ____

seismicity
Includes use on level 2 under seismology(1).

seismology
For general treatments of the subject. See earthquakes for specific treatments. Includes use on level 1 (list A); on level 2 under area terms(1), bibliography(1) and symposia(1). See list B. If level 1, term set options are:
topic [catalogs, core, crust, earthquakes, elastic waves, experimental studies, explosions, interior, mantle, methods, microearthquakes, microseisms, observatories, properties, seismicity, seismic sources, theoretical studies, tsunamis, volcanology]
subtopic [e.g. elastic properties, physical properties, P-waves, S-waves]
—— CR see also earthquakes; engineering geology

selenates
Term introduced in 1981. Includes use on level 2 under minerals(1). See list L.

—— CR see under minerals

selenides
As of 1981, includes use on level 2 under minerals(1). See list L. Before 1981, included use on level 3 under minerals(1) sulfides(2).
—— CR see under sulfides under minerals

selenites
Term introduced in 1981. Includes use on level 2 under minerals(1). See list L.
—— CR see under minerals

selenium
Includes use on level 1 and 2 as chemical element (list D).

Senegal
Formerly a republic in French Community. Achieved independence in 1960. Includes use on level 1 as an area term (list O). For term set options see list B.

sensing see remote sensing

Septibranchia
Term introduced in 1981. Includes use on level 2 under Mollusca(1). See list F.

serpentine group
Includes use in combination with sheet silicates (i.e. sheet silicates, serpentine group) on level 2 under minerals(1). See list L.

settlement
As of 1978, term is used on level 2 under foundations(1), land subsidence(1), and soil mechanics(1).

Seychelles
Level 1 area term as of 1978. Officially designated as the Republic of Seychelles. Island group about 700 miles NE of Malagasy Republic.

shale CR see under ____ under sedimentary rocks
—— see oil shale

shale oil CR see oil shale

shatter cones CR see under ____ under geomorphology; metamorphism

shear zones CR see under ____ under faults

sheet silicates
Includes use on level 2 under minerals(1); in combination with chlorite group, clay minerals, mica group, and serpentine group (i.e. sheet silicates, chlorite group) to form terms on level 2 under minerals(1). See list L.

—— CR see under minerals

shelf see continental shelf

Shetland Islands
Archipelago off N Scotland 50 miles NE of Orkney Islands. Includes use on level 1 as an area term (list O). For term set options see list B.

shoaling
Includes use on level 2 under ocean waves(1).
shock see bow shock waves
shock metamorphism
Term introduced in 1978. Includes use on level 2 under metamorphism(1).
shock waves
Includes use on level 2 under interplanetary space(1); on level 3 under deformation(1) experimental studies(2), and under seismology(1).
shore features
Includes use on level 2 under geomorphology(1).
—— CR see under geomorphology
shorelines
A level 1 term as of 1978. Used for geological studies on the engineering aspects of shorelines. Before 1978, included use on level 2 under engineering geology(1). Term set options are:
barrier islands
  subtopic
beaches
  subtopic
changes
  subtopic
construction
  subtopic
design
  subtopic
dynamics
  subtopic
erosion
  subtopic
hydraulics
  subtopic
management
  subtopic
seawalls
  subtopic
stabilization
  subtopic
—— CR see also under engineering geology under ——
Sierra Leone
Former British colony and protectorate. Became independent in 1961. Includes use on level 1 as an area term (list O). For term set options see list B.
silica minerals
Includes use in combination with framework silicates (i.e. framework silicates, silica minerals) on level 2 under minerals(1). See list L.
silicates
In indexing, use this term only for broad treatments of the entire class of minerals; otherwise, use a narrower term, e.g. orthosilicates, ring silicates, chain silicates, etc. Includes use on level 2 under minerals(1). See list L. As of 1978, the term silicates is autoposted to all individual silicate minerals.

—— CR see under minerals
—— see chain silicates; framework silicates; ring silicates; sheet silicates
silicides
As of 1981, includes use on level 2 under minerals(1). See list L. Before 1981, included use on level 3 under minerals(1) native elements and alloys(2).
—— CR see under alloys under minerals
Silicoflagellata
Includes use on level 2 under Protista(1). See list F.
silicon
Includes use on level 1 and 2 as a chemical element (list D).
sills
Includes use on level 2 under intrusions(1).
—— CR see under intrusions
silt CR see under —— under sediments
Silurian
Includes use on level 1 as an age term (list E); on level 2 under paleoterms, e.g. paleoecology; paleogeography, paleomagnetism. Above Ordovician, below Devonian.
—— CR see also under geochronology under ——; see also under stratigraphy under ——
silver
Includes use on level 1 and 2 as a chemical element (list D). As of 1981, use silver ores for silver as a commodity. Before 1981, included use on level 1 as a commodity term.
silver ores
Term introduced in 1981. Includes use on level 1 and 2 as a commodity term (list C).
—— CR see also under economic geology under ——
simians
Term introduced in 1981. Includes use on level 2 under Mammalia(1). See list F.
Singapore
Island republic and city off the southern tip of the Malay Peninsula. Includes use on level 1 as an area term (list O). For term set options see list B.
Siphonapteroida
Term introduced in 1981. Includes use on level 2 under Insecta(1). See list F.
Sipunculoida
Includes use on level 2 under worms(1). See list F.
Sirenia
Includes use on level 2 under Mammalia(1). See list F.
site exploration
Includes use on level 2 under engineering geology(1).

skarn
Compositional term. As of 1978, includes use on level 2 under metasomatic rocks(1). Before 1978, included use on level 3 under metamorphic rocks(1).
slate deposits
Term introduced in 1981. Includes use on level 1 and 2 as a commodity term (list C).
—— CR see also under economic geology under ——
slates
Includes use on level 2 under metamorphic rocks(1). See list J.
slickensides CR see under —— under lineation
sliding see gravity sliding
slope see continental slope
slope stability
A level 1 term as of 1978. Used for geological studies on the engineering aspects of mass movements. For other aspects, see geomorphology. Before 1978, included use on level 2 under engineering geology(1). Term set options are:
creep
  subtopic
debris flows
  subtopic
earthflows
  subtopic
embankments
  subtopic
erosion
  subtopic
excavations
  subtopic
experimental studies
  subtopic
failures
  subtopic
field studies
  subtopic
landslides
  subtopic
liquefaction
  subtopic
mass movements
  subtopic
mudflows
  subtopic
rockfalls
  subtopic
site exploration
  subtopic
stabilization
  subtopic
talus slopes
  subtopic
theoretical studies
  subtopic
—— CR see also engineering geology; geomorphology; see also under engineering geology under

snow
Includes use on level 2 under hydrology(1); as level 3 under glacial geology(1) glaciers(2).

Society Islands
Includes use on level 1 as an area term (list O). For term set options see list B.

sodalite group
Includes use in combination with framework silicates (i.e. framework silicates,sodalite group) on level 2 under minerals(1). See list L.

sodium
Includes use on level 1 and 2 as a chemical element (list D).

sodium carbonate
Includes use on level 1 as a commodity term (list C).

—— CR see also under economic geology under ——

sodium sulfate
Includes use on level 1 as a commodity term (list C).

—— CR see also under economic geology under ——

soft sediment deformation
Includes use on level 2 under sedimentary structures(1). See list K.

soil erosion
Term introduced in 1982. Includes use on level 2 under soils(1).

soil group
Includes use on level 2 under soils(1). See list M.

soil mechanics
A level 1 term as of 1978. Used for geotechnical studies. Includes use on level 2 under tunnels(1), underground installations(1), and nuclear facilities(1). If 1, term set options are:
analysis
  applications
  subtopic
case studies
  subtopic
concepts
  subtopic
deformation
  subtopic
earth pressure
  subtopic
elasticity
  subtopic
experimental studies
  subtopic
frost action
  subtopic
liquefaction
  subtopic
materials, properties
  subtopic
methods
  subtopic
settlement
  subtopic

site exploration
  subtopic
techniques
  subtopic
theoretical studies
  subtopic

—— CR see also foundations; rock mechanics; underground installations

soils
Includes use on level 1 (list A); on level 2 under area terms(1), education(1), bibliography(1), and symposia(1). Used for general pedology as well as specific topics. If level 1, term set options are:
analysis
  biological methods
  chemical analysis
  physical methods
  sample preparation
biota
  bacteria
  fungi
  Invertebrata
  microorganisms (for several types)
  Protista
  Vertebrata
geochemistry
  ion exchange
  nitrogen
  phosphorus
  (This heading of second rank picks up papers on cation exchange relations, the nitrogen status of soils, transformations of nitrogen, phosphorus fixation, potassium fixation and the like. Listing of nitrogen and phosphorus is illustrative; names of all other elements are allowed.)
classification
  soil group [List M; names of groups in different systems may be used; Paleosols may also be used]
composition
  chemical composition
  mineral composition
  organic materials
conservation
  erosion control
  fertility maintenance
  physical properties
  (This covers maintenance of structure, pore space, good tilth and the like so that root penetration would be facilitated. Included should be papers on deterioration in the physical properties of soils and its prevention.)
correlation
  subtopic
fertilizers
  farms

field studies
  methods or type of methods (e.g. remote sensing, sampling, applications, programs, techniques)
morphology
  color
  horizons
  micromorphology
  soil profiles
  textures
nutrients
  major elements
  toxic substances
  trace elements
organic materials
  abundance
  composition
  distribution
patterns
  subtopic
pedogenesis
  factors or type of factor (e.g. biota, climate, geomorphology, parent materials, time factor)
  horizon differentiation
  leaching
  migration
  organic materials (gains, losses, etc.)
  transformations
properties
  subtopic
salinity
  occurrence
  treatment
soil erosion
  landslides
  water erosion
  wind erosion
soil group
  name of group
surveys (only for detailed study of a particular area)
  area
waste disposal
  animal waste
  human waste
water regimes
  characterization
  drainage
  irrigation
  movement (in soil)
  storage (storage in soil)
yields
  forests
  fruits
  vegetables

solar see astrophysics and solar physics

solar flares
Term introduced on level 2 under astrophysics and solar physics(1) in 1978.

solar system CR see under —— under planetology

solar wind
Includes use on level 2 under interplanetary space(1) and under magnetosphere(1).

Solemyida
Term introduced in 1981. Includes use on level 2 under Mollusca(1). See list F.

solid waste
As of 1978, term is used on level 2 under waste disposal(1).

solifluction
As of 1978, term is used on level 2 under permafrost(1).

—— *CR see under* permafrost; *see under* periglacial features *under* geomorphology; glacial geology

Solomon Islands
Group of islands E of New Guinea in the SW Pacific Ocean. Bougainville, Buka, and Green Islands in the W are part of Papua New Guinea while the remaining 10 large islands and 4 groups of small islands remain a protectorate of the United Kingdom. Includes use on level 1 as an area term (list O). For term set options see list B.

solubility *CR see under* properties *under* geochemistry

solution features
Includes use on level 2 under geomorphology(1) and land subsidence(1).

Somali Republic
Comprises former British Somaliland and Trust Territory of Somalia (formerly Italian Somaliland). Includes use on level 1 as an area term (list O). For term set options see list B.

Somasteroidea
As of 1981, includes use on level 2 under Echinodermata(1). See list F. Before 1981, included use on level 3 under Echinodermata(1) Stelleroidea(2).

sorosilicates
Term introduced in 1981. Includes use on level 2 under minerals(1). See list L.

sorting *CR see under* textures *under* sedimentary rocks; sediments

sources *see* energy sources; heat sources; seismic sources

South Africa
Includes use on level 1 as an area term (list O). For term set options see list B.

South America
Includes use on level 1 or 2 as an area term (list O). If 1, see term set options under list B.

—— *CR see also* Andes; Argentina; Bolivia; Brazil; Chile; Colombia; Ecuador; French Guiana; Guyana; Paraguay; Peru; Surinam; Uruguay; Venezuela

South Australia
Includes use on level 1 as an area term (list O). For term set options see list B.

South Carolina
Includes use on level 1 as an area term (list O). For term set options see list B.

South China Sea
Bounded on N by China and Taiwan, on the E by the Philippine Islands, on the S by Malaysia and on the W by Vietnam. As of 1977, includes use as level 1 area term (list O). For term set options see list B.

South Dakota
Includes use on level 1 as an area term (list O). For term set options see list B.

South-West Africa *CR see* Namibia

Southern Hemisphere
Used when discussing many large areas too numerous to mention. Includes use on level 1 and 2 as an area term (list O). If 1, see list B for term set options.

—— *CR see also* Africa; Antarctic Ocean; Antarctica; Atlantic Ocean; Indian Ocean; Pacific Ocean; South America

Southern U.S.
Term introduced in 1981. Index states as applicable. Includes use on level 1 as an area term. For term set options see List B.

Southern Yemen
Peoples Democratic Republic of Yemen. Includes use on level 1 as an area term (list O). For term set options see list B.

Southwestern U.S.
Term introduced in 1978. Includes use on level 1 as an area term.

Soviet Union *CR see* USSR

space *see* interplanetary space; underground space

Spain
Includes use on level 1 as an area term (list O). For term set options see list B.

Spanish Sahara *CR see* Western Sahara

spectra
As of 1978, term is used on level 2 under minerals(1).

spectrometry *CR see* spectroscopy

spectroscopy
Used for methodology. For data, see under appropriate material. Includes use on level I; on level 3 under chemical element(1) analysis(2). If used on level I, term set options are:
   methods
      name of method [alpha-ray spectroscopy, atomic absorption, electron probe, emission spectroscopy, flame photometry, gamma-ray spectroscopy, infrared spectroscopy, ion probe, laser methods, mass spectroscopy, microwave spectroscopy, Mossbauer spectroscopy, neutron spectroscopy, nuclear magnetic resonance, optical spectroscopy, radio-frequency spectroscopy, Raman spectroscopy, ultraviolet spectroscopy, X-ray spectroscopy, X-ray fluorescence]
   techniques
      topic [e.g. sample preparation]

speleology *CR see under* solution features *under* geomorphology

Spermatophyta
Term introduced in 1981. Includes use on level 1 and 2 as a fossil term (list F).

Sphenopsida
Including Articulatae. Includes use on level 2 under pteridophytes(1). See list F.

spills *see* oil spills

Spiriferida
Includes use on level 2 under Brachiopoda(1). See list F.

Spirillinacea
Includes use on level 2 under foraminifera(1). See list F.

Spitsbergen
Norwegian archipelago, 360 miles N of Norway, including the main island of Spitsbergen plus North East Land, Edge Island, and Barents Island. Part of the Svalbard Island group. Includes use on level 1 as an area term (list O). For term set options see list B.

Spongiae *CR see* Porifera

spreading *see* sea-floor spreading

springs
Includes use on level 1 (list A). Used for papers stressing spring hydrology. Term set options are:
   topic [composition, genesis, geochemistry, hot springs, mineral waters, temperature]
   subtopic (no area term)

—— *CR see also* ground water; thermal waters; *see also under* hydrogeology *under* ____

—— *see* hot springs

Spumellina
Includes use on level 2 under Radiolaria(1). See list F.

Squamata
As of 1981, includes use on level 2 under Reptilia(1). See list F. Before 1981, included use on level 3 under Reptilia(1) Lepidosauria(2).

Sri Lanka
Formerly Ceylon. Includes use on level 1 as an area term (list O). For term set options see list B.

stability
Includes use on level 2 under paleomagnetism(1), foundations(1), land subsidence(1), and tunnels(1).

—— *see* slope stability

stabilization
As of 1978, term is used on level 2 under shorelines(1) and slope stability(1).

standard materials
Level 1 term introduced in 1976. Used for rocks or minerals or other materials that have been designated as standard by geological laboratories. Includes use on level 1 (list A). Term set options are:
  material name (1st level terms only)
  type
  topic [age, analysis, alteration, catalogs, classification, experimental studies, identification, nomenclature, observations, preparation, properties]
  subtopic or name of laboratory

statements *see* impact statements

stocks
Includes use on level 2 under intrusions(1).

Stolonoidea
Includes use on level 2 under Graptolithina(1). See list F.

storage
As of 1978, term is used on level 2 under reservoirs(1).

storms
Includes use on level 2 under meteorology(1) and geologic hazards(1).

—— *see* magnetic storms

stratigraphy
Used for the discipline as a whole. See under age terms (list E). Includes use on level 1 (list A); on level 2 under area terms(1), bibliography(1), education(1), continental shelf(1), and symposia(1). If level 1, term set options are:
  topic [applications, bibliography, catalogs, classification, concepts, education, history, methods, nomenclature, objectives, philosophy, practice, principles, textbooks]
  subtopic

streams *see* rivers and streams

stromatolites
Includes use on level 2 under algae(1); on level 3 under sedimentary structures(1) biogenic structures(2). See list F (fossils) and list K (sedimentary structures).

—— *CR see under* algae

Stromatoporoidea
Includes use on level 2 under Coelenterata(1). See list F.

—— *CR see under* ____ *under* Coelenterata; problematic fossils

strontium
Includes use on level 1 and 2 as a chemical element (list D). As of 1981, use strontium ores for strontium as a commodity. Before 1981, included on level 1 as a commodity term.

strontium ores
Term introduced in 1982. Includes use on level 1 and 2 as a commodity term (list C).

Strophomenida
Includes use on level 2 under Brachiopoda(1). See list F.

structural analysis
Used for the analysis of structural features on a relatively small scale (from thin section to outcrop). The analysis may lead to interpretation of larger-scale features such as folds, fractures, and faults, or tectonics. Includes use on level 1 (list A). Term set options are:
  concepts
    subtopic
  experimental studies
    subtopic
  faults
    subtopic
  folds
    subtopic
  foliation
    subtopic
  fractures
    subtopic
  interpretation
    topic [e.g. axial-plane structures, boudinage, cleavage, elongate minerals, fold axes, folds, foliation, fractures, joints, laminations, layering, linear deformation, lineation, melange, mullions, petrofabrics, planar deformation, preferred orientation, schistosity, slickensides, etc...]
  lineation
    subtopic
  methods
    subtopic [e.g. electron microscopy, universal stage, X-ray analysis]
  preferred orientation
    subtopic
  principles
    subtopic
  theoretical studies
    subtopic
  —— *CR see also* folds; foliation; lineation; tectonics

structural geology
Used for the discipline as a whole. Includes use on level 1 (list A); on level 2 under area terms(1). If 1, term set options are:

topic [applications, bibliography, catalogs, classification, concepts, education, experimental studies, history, instruments, methods, nomenclature, philosophy, practice, principles, textbooks, theoretical studies]
  subtopic

structural petrology *CR see* structural analysis

structure *see* crystal structure

structures
As of 1978, used on level 2 under foundations(1). Term is restricted to engineering geology.

—— *see* biogenic structures; cylindrical structures; planar bedding structures; primary structures; secondary structures; sedimentary structures; turbidity current structures

studies *see* areal studies; case studies; experimental studies; faunal studies; feasibility studies; field studies; floral studies; lunar studies; theoretical studies

style
Includes use on level 2 under folds(1), foliation(1), fractures(1) and lineation(1).

stylolites *CR see under* ____ *under* sedimentary structures

Stylommatophora
Term introduced in 1981. Includes use on level 2 under Mollusca(1). See list F.

Stylophora
Includes use on level 2 under Echinodermata(1). See list F.

subduction
Includes use on level 2 under plate tectonics(1); on level 3 under tectonophysics(1).

submarine canyons
Includes use on level 2 under ocean floors(1).

—— *CR see under* ocean floors

submarine installations
As of 1978, term is used on level 2 under marine installations(1). Term used only on level 3 from 1976 through 1977.

subsidence
As of 1978, includes use on level 2 under neotectonics(1). Also includes use on level 3 under tectonics(1).

—— *see* land subsidence

subways
Term introduced in 1978 on level 2 under tunnels(1).

Sudan
Includes use on level 1 as an area term (list O). For term set options see list B.

Suiformes
Term introduced in 1981. Includes use on level 2 under Mammalia(1). See list F.

sulfate *see* sodium sulfate

sulfates
Includes use on level 2 under minerals(1). See list L.
—— *CR see under* minerals

sulfides
Includes antimonides, arsenides, bismuthides, oxysulfides, selenides, and tellurides. Includes use on level 2 under minerals(1). See list L.
—— *CR see under* minerals

sulfosalts
Includes sulfantimonates, sulfantimonites, sulfarsenates, sulfarsenites, sulfobismuthites, sulfogermanates, sulfostannates, sulfovanadates. Includes use on level 2 under minerals(1). See list L.
—— *CR see under* minerals

sulfur
Includes use on level 1 and 2 as a chemical element (list D). As of 1981, use sulfur deposits for sulfur as a commodity. Before 1981, included use on level 1 as a commodity term.

sulfur deposits
Term introduced in 1981. Includes use on level 1 and 2 as a commodity term (list C).
—— *CR see also under* economic geology *under* ——

sulphur *CR see* sulfur

Sun
Includes use on level 1. See entry under Moon(1) for term set options.

sunspots
As of 1978, term is used on level 2 under geologic hazards(1).

surface phenomena
Includes use on level 2 under astrophysics and solar physics(1).

surface properties
For photographic characteristics, electromagnetic responses, etc. Includes use on level 2 under Moon(1).

surface water
Includes use on level 3 under environmental geology(1). As of 1981, includes use on level 2 under pollution(1).

surfaces *see* erosion surfaces

Surinam
Includes use on level 1 as an area term (list O). For term set options see list B.

survey organizations
Term introduced in 1981. Includes use on level 1 (list A). Used for the work of geological surveys, national or local. Term set options are:
    topic [annual report, current research, history, organization, research]
      name of survey [e.g. U. S. Geological Survey, etc.]

—— *CR see also under* general *under* ——

surveys
As of 1981, restricted to actual surveying and its results. Before 1981, also used on level 1 for the work of geological surveys. See survey organizations. Includes use on level 2 under soils(1), ground water(1), hydrology(1), geochemistry(1), and geodesy(1); and on level 3 in area sets under geophysical surveys(2).

—— *see* acoustical surveys; aeromagnetic surveys; electrical surveys; electromagnetic surveys; geophysical surveys; gravity surveys; infrared surveys; magnetic surveys; magnetotelluric surveys; radioactivity surveys; seismic surveys; telluric surveys

Swaziland
Borders on Mozambique and South Africa. Administered by a British High Commissioner until independence in 1968. Includes use on level 1 as an area term (list O). For term set options see list B.

Sweden
Includes use on level 1 as an area term (list O). For term set options see list B.

Switzerland
Includes use on level 1 as an area term (list O). For term set options see list B.

syenite *CR see under* —— *under* igneous rocks

syenites
Term introduced in 1981. Includes use on level 2 under igneous rocks(1). See list H.
—— *CR see under* igneous rocks
—— *see* alkali syenites

Symmetrodonta
Order. Includes use on level 2 under Mammalia(1).

symposia
Includes use on level 1 (list A); as a level 2 or 3 term appropriate to a large number of topics, e.g. on level 2 under sedimentary petrology(1). See list G. If level 1, term set options are:
    topic [list B or extraterrestrial geology, general, geophysics]
      subtopic

Synapsida
Includes use on level 2 under Reptilia(1). See list F.

synthesis
As of 1978, term is used on level 2 under crystal growth(1).

Syria
Includes use on level 1 as an area term (list O). For term set options see list B.

system *see* solar system

systems
Includes use on level 2 under faults(1), folds(1), and fractures(1).
—— *see* coordinate systems

systems analogs ·
Includes use on level 2 under ground water(1).

Tabulata
Includes use on level 2 under Coelenterata(1). See list F.

Taeniodonta
Includes use on level 2 under Mammalia(1). See list F.

Tahiti
Island of E group of Society Islands in French Polynesia. Includes use on level 1 as an area term (list O). For term set options see list B.

tail *see* magnetic tail

Taiwan
Island off Fuklen Province of China. Seat of Chinese Nationalist government known as the Republic of China. Includes use on level 1 as an area term (list O). For term set options see list B.

talc deposits
Term introduced in 1981. Includes use on level 1 and 2 as a commodity term (list C).
—— *CR see also under* economic geology *under* ——

talus slopes
Term introduced in 1978 on level 2 under slope stability(1).

tantalates
As of 1981, includes use on level 2 under minerals(1). See list L. Before 1981, included use on level 3 under minerals(1) oxides(2).
—— *CR see under* oxides *under* minerals

tantalum
Includes use on level 1 and 2 as a chemical element (list D). As of 1981, use tantalum ores for tantalum as a commodity. Before 1981, included use on level 1 as a commodity term.

tantalum ores
Term introduced in 1981. Includes use on level 1 and 2 as a commodity term (list C).
—— *CR see also under* economic geology *under* ——

Tanzania
Former British U.N. Trust Territory of Tanganyika, which became independent in 1961. United with the island of Zanzibar to become Tanzania in 1964. Includes use on level 1 as an area term (list O). For term set options see list B.

tar sands *CR see* oil sands

Tasman Sea
Between SE Australia and Tasmania on W and New Zealand on the E. As of 1977, includes use as a level 1 area term (list O). For set options see list B.

Tasmania
Island and state S of Victoria. Includes use on level 1 as an area term (list O). For term set options see list B.

Tasmanites
Includes use on level 2 under algae(1). See list F.

taxonomy
Includes use on level 2 under fossil group(1); on level 2 under paleontology(1). See list F.

technetium
Includes use on level 1 and 2 as chemical element (list D).

techniques
This term deals with sampling. Includes use as a level 2 or 3 term appropriate to a large number of topics, e.g. on level 2 under meteorology(1), mineral exploration(1), and spectroscopy(1). See list G.

technology
Includes use on level 2 under mining geology(1).

tectonics
Used for the structural makeup and structural evolution of regions. See also structural analysis. Includes use on level 1. Term set options are:
concepts
　topic
evolution
　(type of structure, e.g. Alpine Orogeny, Appalachian Phase, basement, basin range structure, Caledonian Orogeny, geanticlines, Hercynian Orogeny, lineaments, rift zones, shear zones, tectonic platforms, etc.)
gravity sliding
　topic
vertical movements
　topic (subsidence, uplifts)
—— *CR see also* epeirogeny; faults; folds; geosynclines; neotectonics; orogeny; plate tectonics; salt tectonics; structural analysis; *see also under* structural geology *under* ——
—— *see* global tectonics; plate tectonics; salt tectonics

tectonophysics
Used for treatments of the application of physics in tectonics. Until 1976 this set included documents dealing with plate tectonics(1). Includes use on level 1; on level 2 under area terms(1), bibliography(1), and symposia(1). See list B. If level 1, term set options are:

topic [bibliography, concepts, convection, convection currents, experimental studies, methods, observations, practice, processes, symposia, theoretical studies]
　subtopic

tektites
Includes use on level 1 (list A). Used for the descriptions of small, rounded, pitted, black to green or yellow bodies of silicate glass of nonvolcanic origin. Term set options are:
topic [age, classification, composition, distribution, genesis, mineral composition, properties]
　(type of tektite) or subtopic (no area term)
—— *CR see also* meteorites; *see also under* petrology *under* ——

Teleostei
As of 1981, includes use on level 2 under Pisces(1). Before 1981, included use on level 3 under Pisces(1) Osteichthyes(2).

tellurates
Term introduced in 1981. Includes use on level 2 under minerals(1). See list L.
—— *CR see under* minerals

telluric surveys *CR see* Earth-current surveys *under* geophysical surveys *under* ——

tellurides
As of 1981, includes use on level 2 under minerals(1). See list L. Before 1981, included use on level 3 under minerals(1) sulfides(2).
—— *CR see under* minerals

tellurites
Term introduced in 1981. Includes use on level 2 under minerals(1). See list L.
—— *CR see under* minerals

tellurium
Includes use on level 1 and 2 as a chemical element (list D).

temperature
For temperatures of mineral formations, see phase equilibria(1). Includes use on level 2 under core(1), Earth(1), heat flow(1), lava(1), mantle(1) and meteorology(1); on level 2 under paleoclimatology(1) for paleotemperatures. Also includes use as a level 2 or 3 term appropriate to a large number of topics. See list G.

temperatures *see* densities and temperatures; ion densities and temperatures

Tennessee
Includes use on level 1 as an area term (list O). For term set options see list B.

Tentaculites
Genus. As of 1981, includes use on level 2 under Mollusca(1). See list F. Before 1981, included use on level 3 under problematic fossils(1).

tephrochronology
Includes use on level 2 under geochronology(1).
—— *CR see under* geochronology

terbium
Includes use on level 1 and 2 as a chemical element (list D).

Terebratulida
Includes use on level 2 under Brachiopoda(1). See list F.

Tertiary
Includes use on level 1 as an age term (list E); on level 2 under paleoterms, e.g. paleoecology, paleogeography, paleomagnetism.
—— *CR see also* Eocene; Miocene; Neogene; Oligocene; Paleocene; Paleogene; Pliocene; *see also under* geochronology *under* ——; *see also under* stratigraphy *under* ——

Tethys
As of 1978, term is used on level 2 under continental drift(1). An elongated east-west sea, similar to the Mediterranean Sea, that separated Europe and Africa and extended across southern Asia in Pre-Tertiary time. Index continents as applicable.

Tetrabranchiata
Term introduced in 1981. Includes use on level 2 under Mollusca(1). See list F.

Tetrapoda
As of 1981, includes use on level 1 and 2 as a fossil term (list F). Before 1981, included use on level 3 under Vertebrata(1).

Texas
Includes use on level 1 as an area term (list O). For term set options see list B.

textbooks
Use under disciplines. Includes use as a level 2 or 3 term appropriate to a large number of topics, e.g. on level 2 under economic geology(1), geology(1), and geochemistry(1). See list G.

Textulariina
Includes use on level 2 under foraminifera(1). See list F.

textures
Includes use as a level 2 or 3 term appropriate to a large number of topics, e.g. on level 2 under metamorphic rocks(1). See list G.

Thailand
Includes use on level 1 as an area term (list O). For term set options see list B.

thallium
Includes use on level 1 and 2 as a chemical element (list D).

thallophytes
Includes use on level 1 and 2 as a fossil term (list F).

Thecamoeba
Includes use on level 2 under Protista(1). See list F.

Thecideidina
As of 1981, includes use on level 2 under Brachiopoda(1). Before 1981, included use on level 3 under Brachiopoda(1) Articulata(2).

Thecodontia
Term introduced in 1981. Includes use on level 2 under Reptilia(1). See list F.

theoretical studies
Includes use as a level 2 or 3 term appropriate to a large number of topics, e.g. on level 2 under faults(1), folds(1), fractures(1), heat flow(1), and meteorology(1). See list G.

Therapsida
As of 1981, includes use on level 2 under Reptilia(1). See list F. Before 1981, included use on level 3 under Reptilia(1) Synapsida(2).

Theria
Subclass. As of 1981, includes use on level 2 under Mammalia(1). See list F. Before 1981, included use on level 3 under Mammalia(1).

thermal analysis
Level 1 term introduced in 1978. Used for methodology and not for data. For data, use thermal analysis data. Before 1978, thermogravimetric analysis and differential thermal analysis were used as level 1 terms. Term set options are:
  differential thermal analysis
   subtopic
  methods
   subtopic
  techniques
   subtopic
  thermogravimetric analysis
   subtopic
  thermomagnetic analysis
   subtopic

thermal circulation
As of 1978 term is used on level 2 under ocean circulation(1).

thermal conductivity
Includes use on level 2 under heat flow(1). Introduced as level 2 term in 1976.

—— CR see under heat flow

thermal metamorphism
Not a valid index term through 1977. After 1977, includes use on level 2 under metamorphism(1).

thermal pollution
Term introduced in 1978 on level 2 under pollution(1).

thermal waters
Includes use on level 1 (list A). Term set options are:
  topic [fumaroles, geochemistry, genesis, geysers, temperature]

subtopic
—— CR see also geothermal energy; springs; see also under hydrogeology under ——

thermogravimetric analysis
A valid level 1 term through 1977 used for methods. After 1977, use on level 2 under thermal analysis(1).

—— CR see under thermal analysis

thermohaline circulation
Term introduced in 1978 on level 2 under ocean circulation(1).

thermoluminescence
Includes use on level 2 under geochronology(1); on level 3 under material name or discipline.

—— CR see under geochronology

thermomagnetic analysis
As of 1978, term is used on level 2 under thermal analysis(1).

—— CR see under thermal analysis

thermometry see geologic thermometry

thickness
Includes use on level 2 under crust(1).

thorium
Includes use on level 1 and 2 as a chemical element (list D). As of 1981, use thorium ores for thorium as a commodity. Before 1981, included use on level 1 as a commodity term.

thorium ores
Term introduced in 1981. Includes use on level 1 and 2 as a commodity term (list C).

—— CR see also under economic geology under ——

thrust faults CR see under —— under faults

thulium
Includes use on level 1 and 2 as a chemical element (list D).

Thysanopteroida
Term introduced in 1981. Includes use on level 2 under Insecta(1). See list F.

tides
Includes use on level 2 under aeronomy(1) and under ocean circulation(1).

Tillodontia
Order. Includes use on level 2 under Mammalia(1).

time scales
Includes use on level 2 and 3 under geochronology(1). See list B and list E (age terms).

tin
Includes use on level 1 and 2 as a chemical element (list D). As of 1981, use tin ores for tin as a commodity. Before 1981, included use on level 1 as a commodity term.

tin ores
Term introduced in 1981. Includes use on level 1 and 2 as a commodity term (list C).

—— CR see also under Economic geology under ——

Tintinnidae
Family. Including Calpionellidae. Includes use on level 2 under Protista(1). See list F.

titanium
Includes use on level 1 and 2 as a chemical element (list D). As of 1981, use titanium ores for titanium as a commodity. Before 1981, included use on level 1 as a commodity term.

titanium ores
Term introduced in 1981. Includes use on level 1 and 2 as a commodity term (list C).

—— CR see also under economic geology under ——

Togo
Formerly French Togo. Includes use on level 1 as an area term (list O). For term set options see list B.

Tonga
An archipelago of about 150 islands NE of New Zealand. Formerly a British protectorate which became independent in 1970. Includes use on level 1 as an area term (list O). For term set options see list B.

trace elements CR see under —— under ——

tracers
Includes use on level 2 under isotopes(1).

trachyandesites
Term introduced in 1981. Includes use on level 2 under igneous rocks(1). See list H.

—— CR see under igneous rocks

trachybasalts
Term introduced in 1981. Includes use on level 2 under igneous rocks(1). See list H.

—— CR see under igneous rocks

trachyte CR see under —— under igneous rocks

trachytes
Term introduced in 1981. Includes use on level 2 under igneous rocks(1). See list H.

—— CR see under igneous rocks

tracks CR see under —— under ichnofossils; see under biogenic structures under sedimentary structures

transformations
As of 1978, term is used on level 2 under ocean waves(1). Includes use on level 3 under phase equilibria(1), minerals(1), crystal chemistry(1), or soils(1). Before 1978, transformation was used under ocean waves.

transport
Includes use on level 2 under sedimentation(1).

trapped particles
Includes use on level 2 under magnetosphere(1); on level 3 under magnetosphere(1) magnetic storms(2).

tree rings
Includes use on level 2 under geochronology(1).

trenches
Includes use on level 2 under ocean floors(1); on level 3 under plate tectonics(1).

—— CR see under ocean floors

Trepostomata
Includes use on level 2 under Bryozoa(1). See list F.

Triassic
Includes use on level 1 as an age term (list E); on level 2 under paleoterms, e.g. paleoecology, paleogeography, paleomagnetism. Above Permian (of Paleozoic), below Jurassic.

—— CR see also under geochronology under ____; see also under stratigraphy under ____

Triconodonta
Includes use on level 2 under Mammalia(1). See list F.

Trilobita
Class. Includes use on level 1 and 2 as a fossil term (list F).

trilobites
Term introduced in 1981 as the common name for Trilobita. See list F. To be used for Trilobita when they are used as tools in stratigraphy. Includes use on level 1. If 1, term set options are:
  biostratigraphy
    age [List E]

Trilobitomorpha
Includes use on level 2 under Arthropoda(1). See list F.

Trilophosauria
Term introduced in 1981. Includes use on level 2 under Reptilia(1). See list F.

Trinidad and Tobago
Comprises the islands of Trinidad and Tobago, Atlantic Ocean, off NE coast of Venezuela. As of 1977, includes use as a level 1 area term (list O). For term set options see list B.

tritium
Includes use on level 1 and 2. See list D (chemical elements).

—— CR see also deuterium; hydrogen

troughs
Includes use on level 2 under ocean floors(1).

Trucial Coast CR see United Arab Emirates

tsunamis
Includes use on level 2 under seismology(1).

—— CR see under geologic hazards; seismology; see under catastrophic waves under ocean waves

Tuboidea
Includes use on level 2 under Graptolithina(1). See list F.

Tubulidentata
Includes use on level 2 under Mammalia(1). See list F.

tungstates
Includes use on level 2 under minerals(1). See list L.

—— CR see under minerals

tungsten
Includes use on level 1 and 2 as a chemical element (list D). As of 1981, use tungsten ores for tungsten as a commodity. Before 1981, included use on level 1 as a commodity term.

tungsten ores
Term introduced in 1981. Includes use on level 1 and 2 as a commodity term (list C).

—— CR see also under economic geology under ____

Tunisia
Includes use on level 1 as an area term (list O). For term set options see list B.

tunnels
A valid level 1 term as of 1978. Used for geological studies on man-made tunnels. Before 1978, included use on level 2 under engineering geology(1). Term set options are:
  construction
    subtopic
  design
    subtopic
  excavations
    subtopic
  experimental studies
    subtopic
  feasibility studies
    subtopic
  instruments
    subtopic
  rock mechanics
    subtopic
  seepage
    subtopic
  site exploration
    subtopic
  soil mechanics
    subtopic
  stability
    subtopic
  subways
    subtopic
  theoretical studies
    subtopic

—— CR see also under engineering geology under ____

turbidity current structures
Includes use on level 2 under sedimentary structures(1). See list K.

—— CR see under sedimentary structures

turbulence
Includes use on level 2 under aeronomy(1), meteorology(1) and under ocean circulation(1).

Turkey
Includes all of Turkey within its political boundaries. Includes use on level 1 as an area term (list O). For term set options see list B.

twinning
As of 1978, term is used on level 2 under crystal growth(1).

Tylopoda
Term introduced in 1981. Includes use on level 2 under Mammalia(1). See list F.

Tyrrhenian Sea
Between Corsica and Sardinia on the W, the mainland of Italy on the E, and Sicily on the S. Includes use on level 1 as an area term (list O). For term set options see list B.

Uganda
Former British protectorate which became independent in 1962. Includes use on level 1 as an area term (list O). For term set options see list B.

ultramafics
Term introduced in 1981. Includes use on level 2 under igneous rocks(1). See list H.

ultrametamorphism
Includes use on level 2 under metamorphism(1).

underground installations
A valid level 1 term as of 1978. Used for geological studies on underground cavities (natural or otherwise), excluding tunnels. Before 1978, included use on level 2 under engineering geology(1). Term set options are:
  construction
    subtopic
  design
    subtopic
  excavations
    subtopic
  experimental studies
    subtopic
  feasibility studies
    subtopic
  instruments
    subtopic
  mines
    subtopic
  rock mechanics
    subtopic
  seepage
    subtopic

site exploration
  subtopic
soil mechanics
  subtopic
stability
  subtopic
theoretical studies
  subtopic
underground space
  subtopic
waste disposal
  subtopic
—— *CR see also under* engineering geology *under* ——
underground space
Term introduced in 1978 on level 2 under land use(1) and underground installations(1).
—— *CR see under* —— *under* conservation; land use; underground installations
underground water *CR see* ground water
United Arab Emirates
Includes use on level 1 as an area term (list O). Federation of 7 states which was achieved in 1972. Formerly known as Trucial States, Trucial Oman, or Trucial Coast. Not strictly equivalent to Trucial Coast, but replaces Trucial Coast as first order term. For term set options see list B.
United Kingdom
Index political divisions as applicable. Comprising Great Britain and Northern Ireland. Includes use on level 1 as an area term (list O). For term set options see list B.
—— *CR see also* England; Great Britain; Scotland; Wales
United States
Includes use on level 1 or 2 as an area term (list O). If 1, see term set options under list B.
—— *CR see also* the individual states and regions
uplifts
As of 1978, includes use on level 2 under neotectonics(1). Also includes use on level 3 under tectonics(1).
Upper Volta
Former French protectorate which achieved independence in 1960. Includes use on level 1 as an area term (list O). For term set options see list B.
uranium
Includes use on level 1 and 2 as a chemical element (list D). As of 1981, use uranium ores for uranium as a commodity. Before 1981, included use on level 1 as a commodity term.
uranium ores
Term introduced in 1981. Includes use on level 1 and 2 as a commodity term (list C).

—— *CR see also under* economic geology *under* ——
Uranus
Includes use on level 1 and 2. See entry under Moon(1) for term set options.
urban planning
As of 1978, term is used on level 2 under land use(1).
Uruguay
Includes use on level 1 as an area term (list O). For term set options see list B.
USSR
Includes use on level 1 or 2 as an area term (list O). If 1, see term set options under list B.
Utah
Includes use on level 1 as an area term (list O). For term set options see list B.
vacuum fusion analysis
Term introduced in 1978. Includes use on level 2 under chemical analysis(1).
vanadates
Includes use on level 2 under minerals(1). See list L.
—— *CR see under* minerals
vanadium
Includes use on level 1 and 2 as a chemical element (list D). As of 1981, use vanadium ores for vanadium as a commodity. Before 1981, included use on level 1 as a commodity term.
vanadium ores
Term introduced in 1981. Includes use on level 1 and 2 as a commodity term (list C).
—— *CR see also under* economic geology *under* ——
variations
Includes use on level 2 under magnetosphere(1); on level 3 under Earth(1).
varieties
Includes use as level 2 or 3 term appropriate to a large number of topics. See list G.
varves
As of 1978, includes use on level 2 under geochronology(1). See list K.
—— *CR see* lacustrine features *under* geomorphology; *see under* geochronology; *see under* planar bedding structures *under* sedimentary structures
Venerida
Term introduced in 1981. Includes use on level 2 under Mollusca(1). See list F.
Venezuela
Includes use on level 1 as an area term (list O). For term set options see list B.

Venus
Includes use on level 1 and 2. See entry under Moon(1) for term set options.
vermiculite deposits
Term introduced in 1981. Includes use on level 1 and 2 as a commodity term (list C).
—— *CR see also under* economic geology *under* ——
Vermont
Includes use on level 1 as an area term (list O). For term set options see list B.
Vertebrata
Includes use on level 1 and 2 as a fossil term (list F). Term is to be used only when more specific terms do not apply, or are too numerous to be recorded.
—— *CR see also* Agnatha; Amphibia; Aves; Chordata; coprolites; fossil man; ichnofossils; Mammalia; Pisces; problematic fossils; Reptilia
vertebrates
Term introduced in 1981 as the common name for Vertebrata. See list F. To be used for Vertebrata when they are used as tools in stratigraphy. Includes use on level 1 when more specific common names do not apply or are too numerous to be recorded. If 1, term set options are:
  biostratigraphy
  age [List E]
Victoria
Includes use on level 1 as an area term (list O). For term set options see list B.
Vietnam
North Vietnam (Democratic Republic of Vietnam) and South Vietnam (Republic of Vietnam) were combined into an unified nation on June 24, 1976. Apparently, the new country will be known as the Democratic Republic of Vietnam. Includes use on level 1 as an area term (list O). For term set options see list B.
Virginia
Includes use on level 1 as an area term (list O). For term set options see list B.
viscosity
Includes use on level 2 under magmas(1) and lava(1); on level 3 under deformation(1) field studies(2).
volcanic features
Includes use on level 2 under geomorphology(1).
—— *CR see under* geomorphology
volcanic rocks
Not a valid term from 1975 through 1977. After 1977, includes use on level 2 under igneous rocks(1).

—— CR see under igneous rocks

volcanism
Includes use on level 2 under volcanology(1); on level 3 under igneous rocks(1) volcanic rocks(2) and under lava(1); on level 3 under plate tectonics(1).

—— CR see under volcanology

volcanoes
Used when discussing specific volcanoes. Includes use on level 2 under volcanology(1).

—— CR see under volcanology

—— see mud volcanoes

volcanology
Used for the discipline and specific treatments. Includes use on level 1 (list A); on level 2 under area terms(1), bibliography(1), education(1), and symposia(1). If level 1, term set options are:
topic [applications, bibliography, catalogs, classification, concepts, education, history, instruments, methods, nomenclature, practice, research, symposia, textbooks, theoretical studies]
  subtopic (no area term)
  volcanism [for ancient volcanoes and processes]
  subtopic [causes, classification, concepts, eruptions, fumaroles, processes, volcanoes]
  volcanoes [for recent activity only]
  name of volcano or area

Wales
Includes use on level 1 as an area term (list O). For term set options see list B.

wandering *see* polar wandering

Washington
Includes use on level 1 as an area term (list O). For term set options see list B.

waste *see* industrial waste; liquid waste; radioactive waste; solid waste

waste disposal
A level 1 term as of 1978. Includes use on level 2 under impact statements(1), pollution(1), and underground installations(1). Term set options are:
liquid waste
  topic (effects, experimental studies, impact statements, methods, pollution, storage)
radioactive waste
  subtopic
  seepage
  subtopic
site exploration
  subtopic
solid waste
  subtopic
thermal pollution
  subtopic

topic
subtopic

—— CR see also under engineering geology under ——; see also under environmental geology under ——

water
Includes use on level 2 under meteorology(1), under isotopes(1), and under pollution(1).

—— CR see also ground water; hydrogeology; hydrology; water resources

—— see ground water; pore water; sea water; surface water; underground water

water regimes
Includes use on level 2 under soils(1).

water resources
Includes use on level 1 as a commodity term (list C) for economically oriented papers.

—— CR see also under economic geology under ——

waters *see* artesian waters; connate waters; mineral waters; thermal waters

waterways
A valid level 1 term as of 1978. Used for geological studies on manmade or man-modified water channels. Before 1978, included use on level 2 under engineering geology(1). Term set options are:
canals
  subtopic
channels
  subtopic
design
  subtopic
erosion
  subtopic
floods
  subtopic
harbors
  subtopic
hydraulics
  subtopic
irrigation
  subtopic
rivers and streams
  subtopic
seepage
  subtopic

—— CR see also under engineering geology under ——

waves
Includes use on level 2 under aeronomy(1) and meteorology(1).

—— see bow shock waves; breaking waves; catastrophic waves; elastic waves; electromagnetic waves; ideal waves; internal waves; ocean waves; shock waves

weathering
For treatments emphasizing the process. Includes use on level 1 (list A). See mineral deposits, genesis(1) processes(2). Term set options are:
topic [analysis, classification, environment, experimental studies, rates, textbooks]
  subtopic
type of material [igneous rocks, metamorphic rocks, minerals, sedimentary rocks, sediments]
  specific type (e.g. rock name)

—— CR see also under geochemistry under ——; see also under geomorphology under ——; see also under sedimentary petrology under ——

well-logging
For treatments that stress methodology. Includes use on level 1 (list A). Term set options are:
topic [applications, automatic data processing, instruments, interpretation, methods, techniques]
  subtopic
type [acoustical logging, caliper logging, dipmeter logging, electrical logging, electromagnetic logging, radioactivity]
  elaboration of type

—— CR see also geophysical surveys

Welwitschiales
Term introduced in 1981. Includes use on level 2 under gymnosperms(1). See list F.

West Germany
Officially known as Federal Republic of Germany or Bundesrepublik Deutschland. In W central Europe, bounded on N by North Sea and Denmark, on E by East Germany and Czechoslovakia, on SE by Austria, on S by Austria and Switzerland, on SW by France, and on W by Luxembourg, Belgium and the Netherlands. Introduced as a level 1 area term in 1978.

West Indies
Islands between SE North America and N South America enclosing the Caribbean Sea. Index island groups as applicable. Includes use on level 1 or 2 as an area term (list O). If 1, see term set options under list B.

—— CR see also Bahamas; Barbados; Cuba; Dominican Republic; Greater Antilles; Guadeloupe; Haiti; Jamaica; Lesser Antilles; Puerto Rico; Trinidad and Tobago

West Virginia
Includes use on level 1 as an area term (list O). For term set options see list B.

Western Australia
Includes use on level 1 as an area term (list O). For term set options see list B.

Western Hemisphere
Used when discussing many large areas too numerous to mention. Includes use on level 1 and 2 as an area term (list O). If 1, see list B for term set options.

—— *CR see also* Atlantic Ocean; Central America; North America; Pacific Ocean; South America

Western Interior
Level 1 area term as of 1978. Tremendous region in North America including the Great Plains, the Rocky Mountains, the Basin and Range Province of the U. S., and the interior plateaus of Canada. Index countries as applicable.

Western Sahara
Term introduced in 1981. Includes use on level 1 as an area term (list O). For term set options see list B.

Western U.S.
Term introduced in 1978. Includes use as level 1 area term. As of 1981, term includes five more states: Colorado, Wyoming, Montana, Idaho, and Utah.

whistlers
Includes use on level 2 under magnetosphere(1).

wind *see* solar wind

winds
Includes use on level 2 under aeronomy(1) and under meteorology(1).

Wisconsin
Includes use on level 1 as an area term (list O). For term set options see list B.

wood *see* fossil wood

worms
Includes use on level 1 and 2 as a fossil term (list F).

Wyoming
Includes use on level 1 as an area term (list O). For term set options see list B.

X-ray analysis
Level 1 term introduced in 1978. Used for methodology not data. For data, use X-ray data. X-ray diffraction analysis was used on level 1 through 1977. Term set options are:
  methods
  topic or instruments or application
  techniques
  topic
  X-ray diffraction analysis
  subtopic
  X-ray fluorescence
  subtopic

X-ray radiography
  subtopic
X-ray spectroscopy
  subtopic

X-ray diffraction analysis
A valid level 1 term through 1977 used for methodology. After 1977, used on level 2 under X-ray analysis(1).

—— *CR see under* X-ray analysis

X-ray fluorescence
As of 1978, term is used on level 2 under X-ray analysis(1) for technique or applications.

—— *CR see under* X-ray analysis; *see under* methods *under* chemical analysis

X-ray radiography
As of 1978, term is used on level 2 under X-ray analysis(1).

X-ray spectroscopy
Term introduced in 1978. Includes use on level 2 under X-ray analysis(1) and on level 3 under chemical element(1) analysis(2).

Xenarthra
Term introduced in 1981. Includes use on level 2 under Mammalia(1). See list F.

xenoliths
Includes use on level 2 under inclusions(1).

—— *CR see under* inclusions

xenon
Includes use on level 1 and 2 as a chemical element (list D).

Xenungulata
Term introduced in 1981. Includes use on level 2 under Mammalia(1). See list F.

Yellow Sea
Between NE China and the Korean Peninsula. As of 1977, includes use as level 1 area term (list O). For term set options see list B.

Yemen
Yemen Arab Republic. Includes use on level 1 as an area term (list O). For term set options see list B.

yields
Includes use on level 2 under soils(1).

ytterbium
Includes use on level 1 and 2 as a chemical element (list D).

yttrium
Includes use on level 1 and 2 as a chemical element (list D).

Yugoslavia
Includes use on level 1 as an area term (list O). For term set options see list B.

Yukon Territory
Includes use on level 1 as an area term (list O). For term set options see list B.

Zaire
Formerly Belgian Congo and now officially the Democratic Republic of the Congo. Includes use on level 1 as an area term (list O). For term set options see list B.

Zambia
Formerly Northern Rhodesia. Includes use on level 1 as an area term (list O). For term set options see list B.

zeolite group
Includes use in combination with framework silicates (i.e. framework silicates,zeolite group) on level 2 under minerals(1). See list L.

Zimbabwe
Term introduced in 1981 to replace Rhodesia. Includes use on level 1 as an area term (list O). For term set options see list B.

zinc
Includes use on level 1 and 2 as a chemical element (list D). As of 1981, use zinc ores for zinc as a commodity. Before 1981, included use on level 1 as a commodity term.

zinc ores
Term introduced in 1981. Includes use on level 1 and 2 as a commodity term (list C).

—— *CR see also under* economic geology *under* ——

zircon deposits
Term introduced in 1981. Includes use on level 1 and 2 as a commodity term (list C).

—— *CR see also under* economic geology *under* ——

zirconium
Includes use on level 1 and 2 as a chemical element (list D).

Zoanthiniaria
Includes use on level 2 under Coelenterata(1). See list F.

zoning
Includes use on level 2 under metamorphism(1); on level 3 under metasomatism(1), under crystal growth(1) and under mineral name.

6. SPECIAL LISTS

The special lists which follow consist of categories of terms used as level-one or level-two terms in index sets, as well as classes of first-level terms, e.g., commodity terms (List C), and fossils (List F).

For each list there are explanatory notes on indexing, and/or searching in the Bibliography and Index of Geology (B.I.G.).

Under "Indexing," the current indexing practice is given, consisting of instructions on the construction of index sets, which need to be read along with the instructions under the individual level-one terms involved, in the Alphabetical Term Lists.

The notes on searching attempt to guide the searcher in the use of the list. Searchers might also read the notes on Indexing for further clues but should be aware that these notes reflect current practice which in some cases differs from the past.

The special lists are:

A   Level-one terms

B   Area sets

C   Commodities

D   Elements

E   Geologic age (stratigraphic) terms

F   Fossils

G   General terms

H   Igneous rocks

I   Sedimentary rocks

J   Metamorphic rocks

K   Sedimentary structures

L   Minerals

M   Soils

N   Sediments

O   Geographic terms

A – Level-one terms

This List, which is designed to help in retrospective
searching, consists of terms valid on level-one of the
index sets, during the years indicated, in the following
bibliographies:

B - Bibliography and Index of Geology (1969 to date)

E - Bibliography and Index of Geology Exclusive of North
America (1967-1968)

G - Geophysical Abstracts (1965-1966)

N - Bibliography of North American Geology (1961-1970)

T - Bibliography of Theses in Geology (1965-1966)

These five bibliographies have similar indexing and are
merged in GeoRef. The other four can be used to extend a
search of the Bibliography and Index of Geology back in
time. The GSA Bibliography (E) began in 1933 and the USGS
Bibliography (N) covers literature back to 1785.
Caution: the level-one terms for E and N have not been
analyzed back beyond the years indicated, and may differ in
the earlier volumes.

B.I.G. SEARCHING
This list includes all currently or formerly valid
level-one terms for the index sets in the Subject Index,
together with the year in which the term was established
as a level-one term. For terms established after 1969,
there are often notes on how to search prior to the year
of establishment under the term in the Alphabetical
Term List.

Absolute age   B72-
Absolute age, dates B69-71   E G N T
Absolute age, methods   B69-71
   E G N T
Absorption spectrophotometry   N
Acoustical exploration   N
Acoustical logging   N
Actinium B69-   N
Aden G
Adriatic Sea   B69- E G
Aegean Sea B77-
Aeronomy B75-
Afars and Issas G
Afars and Issas Territory   B73-77
Afghanistan B69-   E G
Africa B69-   E G
Agnatha B81-
Alabama B69-   G N T
Alaska B69-   G N T
Albania B69-   E
Alberta B69-   G N T
Algae B69-   E N T

Algal flora B81-
Algeria B69-   E G
Alkali metals   B69 E
Alluvial fans   N
Alluvial plains   N
Alpha activation analysis   N
Alps B69-   E G
Aluminum B69-   E N T
Aluminum ores   B81-
American Samoa   N
Americium   B69-
Amphibia B69-   E N
Amphibians   B81-
Amphineura   N
Andes B69-   E G T
Andesite   N
Andorra B69-   E
Angiosperms B69-   E N
Angola B69-   E G
Anhydrite   N
Anion exchange-spectrochemical
   analysis   N

Annelida B69-80 E N T
Antarctic Ocean B69-   E G T
Antarctic region G
Antarctica B69-   E G T
Anthozoa B69-71 E N T
Anticlines N
Antimony B69-   E N T
Antimony ores B81
Apennines B69-   E
Appalachian Basin N
Appalachians B69-   G N T
Arabia B69   E G
Arabian Peninsula B69-   E
Arabian Sea B77-
Arachnida B69-71 E N
Archaeocyatha B69-   E N
Archean B78-
Arctic G N
Arctic America N
Arctic Ocean B69-   E G N T
Arctic region B69-   E T
Argentina B69-   E G
Argon B69-   E N T
Arid regions N
Arizona B69-   G N T
Arkansas B69-   G N T
Arsenic B69-   E N
Arsenic ores B81-
Artesian waters and wells B69-71
    E N T
Arthropoda B69-   E N
Artifacts B69-71 E N
Asbestos B69-80 E N
Asbestos deposits B81-
Ascension Island B69-72 E G
Asia B69-   E G T
Asia Minor B69 E
Asphalt N
Associations B69-   E N
Astatine N
Asteroidea B69-71 E N T
Asteroids B70-   G
Asteroyoa B69-71 E N
Astrophysics and solar physics B76-
Atlantic Coastal Plain B70-   N
Atlantic Ocean B69-   E G N T
Atlantic Ocean Islands B81-
Atlantic region B76-
Atmosphere B69-   E G N T
Atomic energy N
Aurora B78-
Australasia B69-   E
Australia B69-   E G
Austria B69-   E G T
Automatic data processing B69-
    E G N T
Autoradiography B70-   N
Aves B69-   E N

Azores B69-   G
Bacteria B69-   E N
Bahamas B69-   G N T
Bahrain B77-
Balearic Islands B74-
Balkan Peninsula B69-   E
Baltic region B69-   E G
Baltic Sea B69-   E G
Banda Sea G
Bangladesh B73-
Barbados B74-   G N
Barents Sea G
Barite B69-80   E N T
Barite deposits B81-
Barium B69-   E N T
Bars N
Basalt N
Base metals B78-   N
Baselevel N
Basin and Range N
Basin and Range Province B78-   G
Basin, structural N
Basins, structural B69-71 E N T
Basutoland B69   E
Batholiths B69-71 E N T
Bauxite B69-   E N T
Bay of Biscay G
Beaches N
Bechuanaland G
Belgium B69-   E G
Belize B75-
Benin B69-   E
Bentonite B69-80 E N
Bentonite deposits B81-
Bering Sea B69-   E G N
Bermuda B69-   G N T
Bermuda Islands G
Beryl N
Beryllium B69-   E G N
Beryllium ores B81-
Bhutan B75-
Bibliography B69-   E G N
Biogeochemical prospecting B69-71 E N
Biogeochemical surveys N
Biogeochemistry N
Biogeography B69-   N
Biography B69-   E N
Birds B81-
Bismark Islands G
Bismuth B69-   E N
Bismuth ores B81-
Bitumens B69-   E N
Bituminous sands N
Black Sea B69-   E G
Blastoidea B69-71   E N
Bogs N
Bohemia B69 E
Bolivia B69-   E G T

Book reviews B73-
Borates N
Borneo B69-73 E G
Boron B69-   E G N T
Boron deposits B81
Botswana B69-   E G
Boudinage B69   E N
Boulders N
Brachiopoda B69-   E N T
Brachiopods B81-
Branchiopoda B69-71 E N
Brazil B69-   E G T
Breccia B69-71 E N T
Brines B69-   E N
British Columbia B69- G N T
British Guiana G
British Honduras B69-74 N T
British Isles G
Bromine B69-   E N
Bromine deposits B81-
Brucite N
Bryophytes B69-   E N
Bryozoa B69-   E N T
Bryozoans B81-
Bulgaria B69-   E G
Burma B69-   E G
Burundi B69-
Cadmium B69-   E N
Calcite B69-80 E N
Calcite deposits B81-
Calcium B69-   E N
Calderas N
Caliche N
California B69-   G N T
Californium B71-   N
Cambodia B69-   E G
Cambrian B69-   E N T
Cameroon B69-   E
Cameroun G
Canada B69-   G N T
Canadian Shield B78-
Canal Zone N
Canary Islands B74-   G
Cape Verde Islands B69-   G
Carbon B69-   E G N T
Carbonate rocks N
Carbonates N
Carboniferous B69-   E N T
Caribbean region B69-   E G N
Caribbean Sea B69-   E G N T
Caroline Islands B69   E
Carpathians B69-   E
Cartography B69-71 E N T
Caspian Sea B69-
Catalogs B69-   E N
Caves B69-71 E N T
Cayman Islands B81-
Celebes Sea B69-   E

Celtic Sea B77
Cenozoic B69-   E N T
Central African Republic B69-   E
Central America B69-   G N T
Cephalopoda B69-71 E N T
Ceramic materials B69-   E N T
Cerium B69-   E N
Cesium B69-   E N T
Ceylon B69-73 E G
Chad B69-   E G
Chad Republic G
Changes of level B69-   E G N T
Chelicerata B69-70 E
Chemical analyses N
Chemical analysis B69-   E N T
Chert N
Chesapeake Bay B69-72 N
Chile B69-   E G T
China B69-   E G
Chlorine B69-   E N T
Chordata B69-   N
Christmas Island B69-71 E
Chromite B69-80 E N T
Chromite ores B81-
Chromium B69-   E N
Cirripedia B69-71 E N
Clay N
Clay mineralogy B69-   E N T
Clays B69-   E N T
Coal B69-   E N T
Coal balls N
Coast Rica N
Cobalt B69-   E N
Cobalt ores B81-
Coelenterata B69-   E N
Collapse structures N
Collections B69-71 E N
Colombia B69-   E G T
Colorado B69-   E G N T
Colorado Plateau B78-   G N
Colorimetric analysis N
Columbia G
Columbia Plateau B78-
Columbia river B69
Columbium B72 N
Comets G
Comoro Islands B77-
Compressibility N
Concretions B69-71 E N
Conglomerate N
Congo B69-   E G
Congo Republic G
Connate water B69-71 N
Connecticut B69-   G N T
Conodonta B81-
Conodonts B69-   E N T
Conservation B78-
Construction materials B69-   E N T

Continental drift   B69-   E G N
Continental margin   G N
Continental shelf   B69-   E N T
Continental slope   B69-   E G N T
Continents   B69-71   E G N
Cook Island   G
Copper   B69-   E N T
Copper ores   B81-
Coprolite   N
Coprolites   B69-   E N
Coral Sea   B77-   G
Corals   B81-
Core   B69-   E G N
Correlation   B69-71   E N
Corsica   B69-   G
Corundum   B69-80   E N
Corundum deposits   B81-
Cosmic dust   B69-71   N
Cosmic-ray methods   G
Cosmogeny   G
Costa Rica   B69-   G N
Cratering   B69-71   E G N
Craters   N
Cretaceous   B69-   E N T
Crete   B69
Crinoidea   B69-71   E N T
Crust   B69-   E G N T
Crustacea   B69-71   E N T
Cryogeology   B69
Cryptoexplosion structures   B69-71
     E G N T
Cryptogam   B69
Crystal chemistry   B69-   E N T
Crystal growth   B71-   N
Crystal structure   B69-   E N
Crystallography   B69-   E N T
Cuba   B69-   G N
Curium   B69-   N
Cyprus   B69-   E G
Cystoidea   B69-71   E N
Czechoslovakia   B69-   E G
Dacite   N
Dahomey   B69-76   E G
Dams   B78-
Deception Island   G
Deformation   B69-   E G N T
Delaware   B69-   G N
Deltas   B69-71   E N T
Denmark   B69-   E G
Density   G
Desert pavement   N
Deserts   N
Deuterium   B69-
Devonian   B69-   E N T
Diabase   N
Diagenesis   B69-   E N T
Diamond   N
Diamonds   B69-   E N
Diapirs   B69-71   E N T

Diatomite   B69-   E N
Diatoms   B69-71   E N
Differential thermal analysis
     B69-77   E N
Dikes   B69-71   E N T
District of Columbia   B69-   G N
Djibouti   B78-
Dolomite   B69-76   E N
Dolomitization   N
Dolostone   B77-80
Dolostone deposits   B81-
Domes   N
Dominican Republic   B72-   N
Drainage changes   N
Drainage patterns   N
Dunes   N
Dysporium   N
Dysprosium   B73-
Earth   B69-   E G N T
Earth current exploration   N
Earth currents   G N
Earth tides   B69-71   E G N
Earth-current exploration   G N
Earth-current methods   B69-71   E G N
Earth-current surveys   B69-71   E G N
Earthquakes   B69-   E G N T
East China Sea   B78-   G
East Germany   B78-
East Pacific Ocean Islands   B81-
Easter Island   B69-72   E
Eastern Hemisphere   B70-
Eastern U. S.   B78-
Echinodermata   B69-   E N T
Echinoderms   B81-
Echinoidea   B69-71   E N T
Ecology   B69-   E N T
Economic geology   B69-   E N T
Ecuador   B69-   E G T
Education   B69-   E G N
Egypt   B69-   E G T
El Salvador   B69-   G N
Elastic properties   B69-71   E G N
Elastic waves   N
Elasticity   N
Elba   G
Electric exploration   N
Electrical exploration   G N
Electrical logging   N
Electrical methods   B69-71   E G N T
Electrical properties   B69-71   E G N T
Electrical surveys   B69-71   E G N T
Electron diffraction analysis   B71   N
Electron-diffraction analysis   N
Electron microscopy   B69-   E N T
Electron paramagnetic resonance   N
Electron probe analysis   N
Electron-probe analysis   N
Electron spin resonance   N
Elements   B69-73   E G N

Emission spectroscopy N
Energy sources B69-   E N
Engineering geology B69-   E N T
England B69-   E G
English Channel B69-   E G
English Channel Islands B81-
Environmental geology B70-
Eocene B71- N
Epeirogenesis B69-71 E N
Epeirogeny B72-
Equatorial Guinea B78-
Erbium B78-
Eritrea G
Erosion B69-71 E N T
Erosion surfaces N
Estuaries B69-71 E N
Ethiopia B69-   E G
Ethiopis G
Eurasia B69-   E
Europe B69-   E G
Europium B69- E N
Eurypterida B69-71 E N
Evaporite deposits B81-
Evaporites B69-80 E N T
Evolution B69-71 E N T
Explosion phenomena B69-71 E G N T
Explosions B78-
Extraterrestrial geology B69-71;
   81- E T
Faeroe Islands B69-   E G
Faeroes G
Falkland Islands B69-   E
Far East B69-   E
Faroe Islands G
Faults B69-   E G N T
Fauna N
Feldspar B69-80 E
Feldspar deposits B81- N
Fennoscandia B69-71 E
Fernando Poo G
Ferns B81-
Fiji B69-   G
Finland B69-   E G
Fish B81-
Fjords N
Flame photometric analysis N
Flame photometry N
Flora B69 N
Florida B69-   G N T
Florida keys B69
Fluid inclusions B69-71; 78- E N
Fluorescence N
Fluorine B69- E N
Fluorite N
Fluorometric analysis N
Fluorspar B69- E N T
Folds B69- E G N T
Foliation B69-   E N T

Foliations N
Foraminifera B69-   E N T
Foraminifers B81-
Formation pressure N
Formosa B69-76 E G
Fossil man B78-
Fossils, problematic B69-77 E N
Fossils, problematical N
Foundations B78-
Fractures B69-   E G N T
France B69-   E G
Francium N
Franz Josef Land G
French Guiana B69-   E
Fuel resources B81-
Fulgurites B69   E N
Fuller's earth N
Fumaroles B69-71 E G N
Fungi B69-   E N
Fusulinidae N
Gabbro N
Gabon B69-   E G
Gadolinium B69- N
Galapagos Islands B69-   E G
Gallium B69-   E N T
Gambia B70-73
Garnet B71 N
Gas chromatographic analyses N
Gas, natural B69-77 E N T
Gastroliths N
Gastropoda B69-71 E N T
Gegenschein G
Gems B69-   E N
General B69-71 E G N
Geobotanical prospecting N T
Geochemical exploration N
Geochemical methods B70
Geochemical prospecting B69-71 E N
Geochemical surveys B69-71 E N T
Geochemistry B69-   E N T
Geochronology B69- E G N T
Geodes N
Geodesy B69-   E G N T
Geologic barometry B69-71 E G N
Geologic cryometry B69 E
Geologic exploration N
Geologic hazards B78-
Geologic mapping N
Geologic thermometry B69-71 E G N
Geological barometry G
Geological exploration B69-71 E N
Geology B72-
Geology as a profession N
Geomorphology B69-   E N T
Geophysical exploration G N
Geophysical logging N
Geophysical methods B69-   E G N T
Geophysical observations B69

Geophysical research G
Geophysical surveys B69-77 E G N T
Geophysics B69- E G N T
Georgia B69- G N T
Geosynclines B69- E G N
Geotectonics N
Geothermal energy B69- E G N
Geothermal gradient G N
Geothermal surveys N
Geothermometry N
Germanium B69- E N
Germany B69- E G
Geysers B69-71 E G N
Ghana B69- E G
Glacial deposits N
Glacial features N
Glacial geology B72- N
Glacial lakes N
Glaciation B69-71 E G N T
Glaciers B69-71 E G N T
Glaciology N
Glauconite B69-80 E N
Glauconite deposits B81-
Glossaries B69- E N
Glossary N
Glossopterio flora B69-71
Goa B69-72 E
Gold B69- E N T
Gold ores B81-
Grabens N
Granite B69-80 E N
Granite deposits B81-
Granodiorite N
Graphite B69-80 E N
Graphite deposits B81-
Graptolites B81- N
Graptolithina B69- E N T
Gravel B69-80 E N
Gravel deposits B81-
Gravity anomalies N
Gravity exploration G N
Gravity field, Earth B69-71 E G N T
Gravity field, Moon B70-71 G
Gravity methods B69-71 E G N T
Gravity surveys B69-71 E G N T
Gravity tectonics N
Great Basin B78-
Great Britain B69- E G
Great Lakes B69-
Great Lakes region B69- G N
Great Plains B78- N
Greater Antilles B78-
Greece B69- E G
Greenland B69- G N T
Greenland Sea G
Groundwater T
Ground water B69- E N T
Guadalupe B69 E

Guadeloupe B69-
Guam G N
Guatemala B69- G N T
Guinea B69- E G
Guinea-Bissau B75-
Gulf Coastal Plain B69- G N
Gulf of Aden B69- E G
Gulf of Alaska G
Gulf of Bothnia G
Gulf of California B69- G N
Gulf of Maine N
Gulf of Mexico B69- G N T
Gulf of Saint Lawrence G N
Gulf of St. Lawrence G N
Guyana B69- E G
Guyots N
Gymnosperms B69- E N
Gypsum B69-80 E N T
Gypsum deposits B81-
Hafnium B69- E N
Haiti B69- N T
Halogens B69-71 E N
Hawaii B69- G N T
Heat flow B69- E G N T
Heat transfer G
Heavy metals N
Heavy minerals B69-80 E N T
Heavy mineral deposits B81-
Helium B69- E N
Helium gas B81-
Hemichordata B76-
High-pressure research N
Highways B78-
Himalaya B69 E
Himalayas B69- E
Historical geology N
History B69-71 E N
Holmium B78-
Holocene B71-
Holothuroidea B69-71 E N
Honduras B71- N
Hong Kong B69- E
Hungary B69- E G
Hydrocarbons N
Hydrogen B69- E N
Hydrogeology B69- E N T
Hydrology B75-
Hydrosphere N
Hydrothermal alteration B69-71 E N T
Hydrothermal solutions N
Hydrozoa B69-71 E N
Icarus G
Ice N
Ice ages (ancient) N
Ice ages, ancient B69-71 E N
Ice islands N
Ice, non-glacial B69-71 E N T
Ice, nonglacial G N

Iceland B69-   E G
Ichnofossils B72-
Idaho B69- G N T
Igneous petrology N
Igneous rocks B69-   E N T
Illinois B69-   G N T
Illinois Basin N
Ilmenite N
Impact phenomena G N
Impact statements   B78-
Impactite N
Impactites G N
Inclusions B69-   E N T
India B69-   E G T
Indian Ocean B69-   E G T
Indian Ocean Islands   B81-
Indiana B69-   G N T
Indium B69-   E N
Indochina B69-   E
Indonesia B69-   E G T
Industrial materials B68
Industrial minerals B69-   E N T
Infrared exploration G N
Infrared methods B69-71   E G N
Infrared spectroscopy N
Infrared surveys B69-71   G N
Insecta B69-   E N
Interplanetary space B75-
Intrusions B69- E N T
Invertebrata B69-   E N T
Invertebrates B81-
Iodine B69-   E N
Iodine deposits B81-
Ionian Sea B69-   E
Ionosphere B77-
Iowa B69-   G N T
Iran B69-   E G
Iraq B69-   E G
Ireland B69-   E G
Iridium B69-   E N
Irish Sea B69-   E G
Iron B69-   E N T
Iron ores B81-
Island arcs N
Isostasy B69-   E G N
Isotopes B69-   E G N T
Israel B69-   E G
Italy B69-   E G
Ivory Coast B69-   E G
Jamaica B69-   G N
Jan Mayen B69-   E G
Japan   B69-   E G
Japan Sea B69-   E G
Johnston Island N
Joints G N
Jordan B69-   E G
Jupiter B69-   G
Jurassic B69-   E N

Kansas B69-   G N T
Kaolin B69-80 E N
Kaolin deposits B81-
Karst N
Kashmir B69-71
Katanga B69-73
Kentucky B69-   G N T
Kenya B69-   E G
Kerguelen Islands G
Kermadec Islands G
Korea B69-   E G
Krypton B69-   E G N
Kuril Islands B69 E
Kuwait B69-   E
Labrador B69-   G N T
Laccoliths N
Lake Erie G
Lake Ontario G
Lake Superior. G T
Lake Superior region B69-70 G N
Lakes B69-71 E N T
Lakes, extinct B69-71 E N T
Lamprophyre N
Land subsidence B78-
Land use B78-
Landforms N
Landslides N
Lanthanides B69-72 E
Lanthanum B69-   E N
Laos B69-   E
Lapland B69
Laser methods G N
Laser surveys N
Laterite N
Laterites B69-71 E N
Lava B69-   E N T
Lava flows N
Lead B69-   E G N T
Lead ores B81-
Lead-zinc deposits B78-
Lebanon B69-   E G T
Lesotho B69-
Lesser Antilles B77-
Leveling G
Leveling Networks G
Lexicon N
Lexicons B69-   E N
Liberia B69-   E G
Libya B69-   E G
Lichens B70-
Life G
Lignite B69-   E N T
Limestone B69-80 E N T
Limestone deposits B81-
Lineaments N
Lineation B69-   E N T
Lithium B69-   E G N
Lithium ores B81-

Lithofacies N
Loess N
Louisiana B69-   G N T
Low-temperature analysis N
Luminescence B69-71   E G N T
Lutetium B69-
Luxembourg B69-   E G
Lycopoda B69-71
Macao G
Macedonia B69
Macquarie Island G
Madagascar B69-74   E G T
Madeira B77-   G
Magma G
Magmas B69-   E G N T
Magmas and magmatic differentiation
   N
Magnesite B69-80   E N
Magnesite deposits B81-
Magnesium B69-   E N
Magnesium ores B81-
Magnetic anomalies N
Magnetic exploration G N
Magnetic field of the Earth N
Magnetic field, Earth B69-71
   E G N T
Magnetic field, Jupiter G
Magnetic field, Mars B69 G
Magnetic field, Mercury B69 G
Magnetic field, Moon B69-71 G
Magnetic field, Sun G
Magnetic field, Venus B69 G
Magnetic field, asteroids B69
Magnetic field, interplanetary G
Magnetic field, planets G
Magnetic methods B69-71   E G N T
Magnetic properties B69-71   E G N T
Magnetic surveys B69-71   E G N T
Magnetite B71 N
Magnetosphere B75-
Magnetotelluric exploration G N
Magnetotelluric methods B69-71
   E G N T
Magnetotelluric surveys B69-71
   E G N T
Maine B69-   G N T
Major-element analyses B69-71
   E N T
Malacostraca B69-71 E N
Malagasy Republic B74-
Malawi B69-   E G
Malay Archipelago B71-
Malaya B69-72   E G
Malaysia B69-   E G
Maldive Islands B69-   E
Mali B69-   E G
Malta B73-
Mammalia B69-   E N T

Mammals B81-
Man, fossil B69-77 E N
Manganese B69-   E N T
Manganese ores B81-
Manitoba B69-   E G N T
Mantle B69-   E G N T
Maps B72-
Marble B69-80   E N
Marble deposits B81-
Mariana Islands B69-   E
Marine geology B69-   E G N T
Marine installations B78-
Maritime Provinces   B78-
Mars B69-   E G
Marshall Islands B69-   E
Martinique N
Maryland B69-   G N T
Mass spectroscopy N
Mass wastage N
Massachusetts B69-   G N T
Mathematical geology B71-
Mauritania B69-   E G T
Mauritius B76-   G
Medical geology B71
Mediterranean region B69-   E G
Mediterranean Sea B69-   E G T
Melanesia B74-
Mercury B69-   E G N T
Mercury ores B81-
Mercury Planet B75-
Merostomata B69-70   E N
Mesozoic B69-   E N T
Metal B69 E
Metals B69-   E N T
Metal ores B81-
Metamorphic petrology N
Metamorphic rocks B69-   N T
Metamorphism B69-   E N T
Metasomatic rocks   B78-
Metasomatism B69-   N T
Metazoa B69
Meteor craters B69-   E G N
Meteorites B69-   E G N T
Meteorology B75-
Meteors G
Mexico B69-   G N T
Mica B69-80   E N
Mica deposits B81-
Michigan B69-   G N T
Micronesia B69-   E
Micropaleontology B69-   E N T
Microscope methods B69-71   E N
Microscope techniques N
Microscopic methods B69   E
Microseisms B69-71   E G N T
Microwave exploration N
Microwave methods G N
Microwave surveys G N

Middle East B69- E G
Midway Islands N
Midwest B78-
Migmatites N
Military geology B69-70 N
Mineragraphy B69-71 E N
Mineral collecting B69-71 E N
Mineral collections N
Mineral data B69-71 E N T
Mineral deposits, genesis B69-
  E N T
Mineral descriptions N
Mineral economics B69-71 E N T
Mineral exploration B69- E G N T
Mineral resources B69- E N T
Mineral zoning B69-71 E N
Mineralogy B69- E N T
Minerals B71-
Mining geology B69- E N T
Minnesota B69- G N T
Minor-element analyses N
Miocene B69- E N
Mississippi B69- G N T
Mississippi Delta G
Mississippi River N
Mississippi Valley B78- G N
Mississippi embayment N
Mississippian B69- N T
Missouri B69- G N T
Models N
Mohorovicic discontinuity B69-
  E G N
Mollusca B69- E N T
Mollusks B81-
Molybdenum B69- E N T
Molybdenum ores B81-
Monaco B69- E
Monazite B69-80 E N
Monazite deposits B81-
Mongolia B69- E G T
Montana B69- G N T
Moon B69- E G N T
Morocco B69- E G
Mossbauer analysis N
Mossbauer effect G
Mozambique B69- E G
Mud volcanoes B69- E N
Mudflows N
Museums B69- E N
Myriapoda B69-71 E N
Namibia B81-
Natural bridges N
Natural gas B78-
Near East B69 E
Nebraska B69- G N T
Neodymium B71-
Neogene B78-
Neon B69- E G N

Neotectonics B78-
Nepal B69- E
Nepheline syenite N
Neptune B69- G
Neptunium B72-
Netherlands B69- E G
Netherlands Antilles B69- N
Neutron activation analysis N
Neutron diffraction analysis N
Nevada B69- G N T
New Britain G
New Britain Island G
New Brunswick B69- G N T
New Caledonia B69- E G
New England B69- G N
New Guinea B69- E G T
New Hampshire B69- G N T
New Hebrides B69- E G
New Jersey B69- G N T
New Mexico B69- G N T
New South Wales B69- E
New York B69- G N T
New Zealand B69- E G T
Newfoundland B69- G N T
Nicaragua B70- G N
Nickel B69- E N T
Nickel ores B81-
Niger B69- E
Niger Republic G
Nigeria B69- E G
Niobium B69- E N T
Niobium ores B81-
Nitrate deposits B81-
Nitrates B68-80 N
Nitrogen B69- E N
Noble gases B71- N
Nodules B69- E N T
Nonmetal deposits B81-
Nonmetals B74-
North Africa B69-71 E
North America B69- G N T
North Carolina B69- G N T
North Dakota B69- G N T
North Korea G
North Sea B69- E G T
North Vietnam G
Northern Hemisphere B69-
Northern Ireland B69- E G T
Northern Territory B69- E
Northwest Territories B69- G N T
Norway B69- E G
Norwegian Sea G
Nova Scotia B69- G N T
Nuclear explosions B69-71 E G N T
Nuclear facilities B78-
Nuclear magnetic resonance N
Nuclear science N
Nucleosynthesis G

Nutation G
Ocean basins B72-   G N
Ocean circulation   B78-
Ocean floors B72-
Ocean waves B72-
Oceania B69- E
Oceanography B69- EG N T
Oceans G N
Oceans, circulation   B72-77
Ohio B69-   G N T
Oil and gas fields B69-   E N T
Oil sand B70-71
Oil sands B69; 72-   N
Oil shale B69-   E N T
Okhotsk Sea B69-   E G
Oklahoma B69-   G N T
Oligocene B69-   E
Oman B69-   E
Ontario B69-   G N T
Oolites N
Optical data processing N
Optical mineralogy B69-71 N T
Optical properties B69 E
Ordovician B69- E N T
Oregon B69-   G N T
Organic materials B69-   E N T
Orogeny B69-   E G N T
Osmium B69-   E N
Ostracoda B69-   E N T
Ostracods B81-
Oxygen B69-   E N T
Ozone B73
Pacific Coast B78-
Pacific Islands G N
Pacific Ocean B69-   E G N T
Pacific region B69-
Pakistan B69-   E G T
Paladium N
Paleobiogeography N
Paleobotany B69- EN T
Paleocene B71-
Paleoclimatology B69-   E G N T
Paleoecology B69-   E N T
Paleogene B78-
Paleogeography B69-   E N T
Paleomagnetism B69-   E G N T
Paleontology B69-   E N T
Paleosalinity G N
Paleotemperature G
Paleotemperatures N
Paleozoic B69- E N T
Palladium B69-   E N
Palynology B69-   E N
Palynomorphs B69-   E N T
Panama B69-   G N T
Pantellerite N
Paper chromatography N
Papua B69-75 E G

Papua New Guinea B75-
Paragenesis B69-   E N T
Paraguay B69-   E
Patterned ground B69-71 E N
Peat B69-   E N
Pebbles B69-71 E N
Pediments N
Pegmatite B69-   E N T
Pegmatites N
Pelecypoda B69-71 E N T
Peneplanes N
Pennsylvania B69-   G N T
Pennsylvanian B69-   N T
Peridotite N
Periglacial features N
Periglacial phenomena N
Perlite N
Permafrost B69-71; 78- E G N
Permeability B69-71 E G N
Permian B69-   E G N T
Persian Gulf B69-   E G
Peru B69- E G T
Petrofabrics B69-71 E G N T
Petrogenesis N
Petrography B69   E N
Petroleum B69-   E N T
Petroleum engineering N
Petrology B69-   E N T
Phanerozoic B72- N
Phase equilibria B69-   E G N T
Philippine Islands B69-   E T
Philippine Sea B78-   G
Philippines G
Phosphate B69-80 E N T
Phosphate deposits B81-
Phosphorescence N
Phosphorous N
Phosphorus B69-   E N
Photogeology B69-71 E G N T
Photomicroscopy N
Physical geology N
Physical properties G N
Phytoliths B69
Pisces B69-   E N T
Placers B69-   E N
Planetology B71-
Planets B69-71 E G
Plantae B72-
Plasticity N
Plate tectonics B76-
Platinum B69-   E N
Platinum ores B81-
Playas N
Pleistocene B71-   N
Pliocene B71-
Pluto B69-   G
Plutonium B69-   E G N
Plutons N

Poland B69-   E G
Polar wandering G
Polarographic analysis N
Pollution B78-
Polonium B71- N
Polymetallic ores B69-   E N
Polynesia B74-
Popular and elementary geology
    B69-72 E N
Porifera B69- E N
Porosity B69-71 E G N T
Portugal B69-   E G
Portuguese Guinea B69-73   E
Portuguese Timor B76
Potash B69-   E N
Potassium B69-   E N T
Praesodymium B78-80
Praseodymium B81-
Precambrian B69-   E N T
Primates N
Prince Edward Island B70-   G N
Problematic fossils B78-
Problematical fossils N
Promethium B72-
Protactinium B69-   N
Proterozoic B69; 78- E
Protista B69-   E N T
Protozoa B69-71   E N
Pseudomorphs N
Pteridophytes B69-   E N
Pterobranchia B70-   N
Pteropoda B69-71   E N
Puerto Rico B69-   G N T
Pumice B70-80 N
Pumice deposits B81-
Pyrenees B69-   E
Pyrite B69-80   E N
Pyrite ores B81-
Qatar B74-

Quark G
Quartz crystal B69-   E N
Quartzite N
Quaternary B69-   E N T
Quebec B69-   G N T
Queensland B69-   E
Radar exploration N
Radar methods G N
Radar surveys G N
Radioactive-waste disposal N
Radioactivity B69-71 E G N T
Radioactivity exploration G N
Radioactivity logging N
Radioactivity methods B69-71
    E G N T
Radioactivity surveys B69-71
    E G N T
Radiochemical analysis N
Radiolaria B69-   E N T

Radiolarians B81-
Radiowave methods N
Radiowave surveys G N
Radium B69-   E N
Radon B69-   E N
Rare earth deposits B81-
Rare earths B69-   E N T
Rare gases B71 N
Reclamation B78-
Red Sea B69-   E G T
Red Sea region B76-
Reefs B69-   E N T
Reflectivity N
Refractometry N
Refractory materials N
Remote sensing B78-
Remote-sensing methods G N
Remote-sensing surveys G N
Reptiles B81-
Reptilia B69-   E N
Reservoirs B78-
Reunion B69-   E
Reunion Island G
Rhenium B69-   E N
Rhode Island B69-   G N T
Rhodesia B69-80 E G
Rhodium B69-   N
Ripple marks N
Rivers B69-71 E N T
Rock creep N
Rock glaciers N
Rock mechanics B78-   G N
Rocky Mountains B69-   G N
Romania B69-   E
Rubidium B69-   E N T
Rumania G
Ruthenium B69-   N T
Rwanda B69-   E G
Ryukyu Islands B69-68 E
Sahara B69-   E
Saint Helena G
Saint Pierre and Miquelon B81-
Saint-Pierre and Miquelon N
Salt B69-   E N T
Salt tectonics B69-   E G N T
Salt water intrusion N
Samarium B71-   N
Samoa B69- E T
Samoa Islands B69   E
Sampling N
San Marino B81-
Sand B69-80   E N
Sand volcanoes B69   E
Sands B81-
Sandstone B69-80 E N
Sandstone deposits B81-
Sarawak B69-75 E
Sardinia B74-   G

Sun B69- G
Surinam B69- E G
Survey organizations B81-
Surveys B69-80 E N T
Swaziland B69- E G
Sweden B69- E G
Switzerland B69- E G
Symposia B69- E N
Syria B69- E G
Tahiti B69; 71- E
Taiwan B77- G
Talc B69-80 E N T
Talc deposits B81-
Tanganyika B69 E
Tantalum B69- E N T
Tantalum ores B81-
Tanzania B69- E G
Tasman Sea B77- G
Tasmania B69- E G T
Technetium B76-
Tectites G
Tectonics B69- E G N T
Tectonophysics B71-
Tektites B69- E G N T
Television logging N
Tellurium B69- E N
Temperature methods G
Temperature surveys G
Tenerife G
Tennessee B69- E G N T
Terbium B70-72; 78-
Terraces N
Tertiary B69- E N T
Tetrapoda B81-
Texas B69- G N T
Thailand B69- E G
Thallium B69- E N
Thallophytes B71-
Thermal analysis B78-
Thermal conductivity G N
Thermal methods G N
Thermal properties G N
Thermal springs B69 E G N
Thermal surveys G N
Thermal waters B69- E N
Thermodynamic properties B69-73
  E G N
Thermogravimetric analysis
  B73-77 N
Thermoluminescence G N
Thorium B69- E N T
Thorium ores B81-
Thoron B69
Throium B70
Thulium B71-72; 78- N
Till N
Tillites N
Timor B69-73 E

Timor Sea G
Tin B69- E N T
Tin ores B81-
Titanium B69- E N T
Titanium ores B81-
Togo B69- E
Tonga B69- E
Tonga Islands B69 E G
Trace elements N
Trace-element analyses B69-71
  E N T
Tracks and trails B69-71 E N T
Traps N
Triassic B69- E N T
Trieste B69-71 E
Trilobita B69- E N T
Trilobites B81-
Trinidad B69-76 G N
Trinidad and Tobago B77-
Tristan da Cunha G
Tritium B69- E N
Trucial Coast B69-73 E
Trucial Oman B69-72
Tsunami G
Tsunamis B69-71 E G N T
Tuff N
Tungsten B69- E N T
Tungsten ores B81-
Tunisia B69- E G
Tunnels B78-
Turbidity currents N
Turkey B69- E G
Tyrrhenian Sea B69- E
U.S.S.R. G
Uganda B69- E G
Ultramafic rocks N
Ultraviolet spectroscopy N
Unconformities B69-71 E N T
Underground installations B78-
Uniformitarianism N
United Arab Emirates B74-
United Kingdom B74- G
United States B69- G N T
Uplifts B69-71 E N
Upper Volta B69- E G
Uranium B69- E N T
Uranium ores B81-
Uranus B70- G
Uruguay B69- E G
USSR B69- E
Utah B69- G N T
Valleys N
Vanadium B69- E N
Vanadium ores B81-
Varves N
Veins B69-71 E N T
Venezuela B69- E G T
Venus B69- E G

Vermes   B69; 74
Vermiculite   B69-80   E N
Vermiculite deposits   B81-
Vermont   B69-   G N T
Vertebrata   B69-   E N T
Vertebrates   B81-
Victoria   B69-   E
Viet Nam   G
Vietnam   B69-   E G T
Virgin Islands   G N T
Virginia   B69-   G N T
Volcanism   B69-71   E G N T
Volcanoes   B69-71   E G N T
Volcanology   B72-
Wake Island   N
Wales   B69-   E G
Washington   B69-   G N T
Waste disposal   B78-
Water   B69   E N
Water falls   N
Water resources   B71-
Waterways   B78-
Weathering   B69-   E N T
Well logging   G N
Well-logging B69-   E T
Wells and drill holes   B69-71
   E N T
West Africa   B69-72   G
West Germany   B78-
West Indies   B69-   G N
West Virginia   B69-   G N T
Western Australia   B69-   E T
Western Hemisphere B69-
Western Interior   B78-

Western Sahara   B81-
Western U.S.   B78-
Williston Basin   N
Windward Islands   T
Wind work   N
Wisconsin   B69-   G N T
World   B69   E N
Worms   B69-   E N
Wyoming   B69-   G N T
X-ray analysis   B78-
X-ray diffraction   N
X-ray diffraction analyses   N
X-ray diffraction analysis
   B69-77   E N T
X-ray fluorescence analysis   N
X-ray radiography   N
X-ray spectrographic analysis   N
Xenoliths   N
Xenon   B69-   E G N
Yellow Sea   B78-   G
Yemen   B69-   E
Ytterbium   B70; 77-   N
Yttrium   B69-   E N
Yugoslavia   B69-   E G
Yukon   G N T
Yukon Territory   B69-
Zaire   B72-
Zambia   B69-   E G T
Zimbabwe   B81-
Zinc   B69-   E N T
Zinc ores   B81-
Zircon   B69-80   E N
Zircon deposits   B81-
Zirconium   B69-   E N

B - Area sets

Index entries for areas in the Bibliography and Index of
Geology consist of an area term which is valid on the first-
level (see List 0), followed by one of the second- and third-
level combinations in this List.

---

areal geology (2)
     Used for entries that might properly be placed under
     three or more of the second-level headings such as
     geomorphology, stratigraphy, structural geology.
     topic (3)
          bibliography, explanatory text (for a map), guidebook,
          maps or locality or regional.

          e.g. France
                    areal geology
                    Narbonne

economic geology (2)

     commodity (List C) (3)

          e.g. Germany
                    economic geology
                    coal

engineering geology (2)
     topic (3)
          See Main List, second-level terms or dams, earthquakes,
          foundations, geologic hazards, highways, land sub-
          sidence, marine installations, nuclear facilities,
          permafrost, reservoirs, rock mechanics, shorelines,
          slope stability, soil mechanics, tunnels, underground
          installations, waste disposal, waterways.
          e.g. California
                    engineering geology
                    earthquakes

environmental geology (2)
     topic (3)
          See Main List, second-level terms, or conservation,
          geologic hazards, impact statements, land use,
          pollution, reclamation, waste disposal.
          e.g. Idaho
                    environmental geology
                    land use

general (2)

     topic (3)
          E.g. annual report, bibliography, current research,
          education, geology, maps, philosophy, research
          e.g. United States
                    general
                    current research

geochemistry (2)

   isotopes or element or other terms from List D or
material or topic (Main List) (3)

     e.g. USSR
         geochemistry
           isotopes

geochronology (2)

   topic (3)
     absolute age or topic from Main List, second-level
     terms or age (List E).

     e.g. United States   e.g. Maine
         geochronology   geochronology
           absolute age   Precambrian

geomorphology (2)

   topic (3)
     See Main List, second-level terms, or glacial geology,
     meteor craters, changes of level, weathering.

     e.g. Sahara
         geomorphology
           eolian features

geophysical surveys (2)

   topic (3)
     Type of survey (see Main List under geophysical
     surveys) or geodesy, heat flow, remote sensing, well-
     logging.

     e.g. Oklahoma
         geophysical surveys
           infrared surveys

hydrogeology (2)

   topic (3)
     ground water, hydrology, springs, thermal waters

     e.g. Louisiana
         hydrogeology
           ground water

mineralogy (2)

   Name of mineral group (List L). In the case of miscellaneous minerals, that term suffices here.

     e.g. Italy
         mineralogy
           framework silicates, feldspar group

oceanography (2)

topic (3)
See Main List, second-level terms; or continental shelf, continental slope, estuaries, geophysical methods, geophysical surveys, marine geology, nodules, ocean basins, ocean circulation, ocean floors, ocean waves, reefs, sea ice, sea water, sedimentation, sediments or locality.

e.g. Pacific Ocean
        oceanography
            marine geology

paleobotany (2)

fossil group (List F, first-level terms) (3)
Repeat set if more than one equally-emphasized fossil group.

e.g. Arizona
        paleobotany
            gymnosperms

paleontology (2)
Also used for studies on both fauna and flora.

Fossil group (List F, first-level terms) (3)
Repeat set if more than one equally-emphasized fossil group.

e.g. California
        paleontology
            foraminifera

petrology (2)

topic (3)
fluid inclusions, igneous rocks, inclusions, intrusions, lava, magmas, metamorphic rocks, metamorphism, metasomatism, meteorites, phase equilibria, tektites, volcanism, volcanology.

e.g. Colorado
        petrology
            metamorphism

sedimentary petrology (2)

topic (3)
clay mineralogy, diagenesis, heavy minerals, reefs, sedimentary rocks, sedimentary structures, sedimentation, sediments, weathering.

e.g. New South Wales
        sedimentary petrology
            sedimentary rocks

seismology (2)

topic (3)
See Main List, second-level terms.

e.g. Washington
          seismology
              earthquakes

soils (2)

topic (3)
See Main List, second-level terms.

e.g. Alberta
          soils
              name of soil group

stratigraphy (2)

age (List E) or
archaeology, biogeography, changes of level, continental drift, paleogeography, paleomagnetism,
paleoclimatology. (3)

Repeat set if more than one equally-emphasized age.

e.g. South Carolina
          stratigraphy
              Cretaceous

structural geology (2)

topic (3)
deformation, epeirogeny, faults, folds, foliation,
fractures, geosynclines, lineation, orogeny,
neotectonics, salt tectonics, structural analysis,
tectonics.

e.g. India
          structural geology
              tectonics

tectonophysics (2)

topic (3)
See Main List, second-level terms; or continental
drift, core, crust, heat flow, mantle, plate tectonics,
sea-floor spreading.

e.g. North America
          tectonophysics
              plate tectonics

volcanology (2)
For Quaternary volcanic activity.

Name of volcano, or locality for several volcanoes (3)

e.g. Italy
          volcanology
              Mount Etna

C – Commodities

The terms in this list are used for earth materials of
actual or potential commercial value, i.e. from the view-
point of economics.

Beginning in 1981, for some commodities, new terms were
adopted to distinguish between the economic and geochemical
viewpoints. For example, beryllium ores was added, to be
used for the economic, and beryllium, the existing term, was
restricted to the geochemical sense. Where new terms have
been added, this is noted below in the list and in the body
of the Term List. Prior to 1981, these commodity terms
were also used for geochemical treatments.

INDEXING
   Commodities designated as (1,2,3) are used on level three
   under area terms (1) and economic geology (2). Each such
   term set causes a cross reference to be generated in the
   Subject Index of the B.I.G., from the commodity to the
   area term.

      e.g. Wyoming   coal see also under
              economic geology   economic geology
                 coal   under Wyoming

   Commodities are also used as first-level terms when no
   geography is emphasized. In this case the set options
   are:

   commodity (1)
      topic (2)
         Such as affinities, economics, exploration, genesis,
         geochemistry, inventory, possibilities, production,
         properties, reserves, resources

         subtopic (3)
            Such as one of the topics listed above.

            e.g. gold ores
                    reserves
                    possibilities

In general, only the most specific commodity term applicable should be used. However, a general term like heavy
mineral deposits or mineral resources can be used with
one or two of the specific commodities which it includes
where these commodities have received special attention
in a discussion of the group.

B.I.G. SEARCHING
   Most papers on commodities deal with the commodities in
   relation to geographic locations. To find commodity
   papers on an area, look under the area (1), economic
   geology (2), and the commodity (3). Another approach,
   which will show the areas related to a particular
   commodity, is to look under the commodity. There, see
   also references to each of these areas will appear.

e.g.   Coal see also under economic geology under
       Alabama; Alberta; Australia; Colorado;
       England; etc.

For those papers on commodities which are unrelated to
specific areas, look under the commodity. For papers on
the genesis of commodities look under mineral deposits,
genesis (1).

Other sets related to commodities can be found under
sedimentary rocks (for coal), metamorphic rocks (for
slate deposits or marble deposits), and igneous rocks
(for granite deposits).

_____

In the list which follows, the commodities are used in index
sets on all three levels, except those followed by (3) which
are only permitted in sets on level three.
Asterisks indicate terms introduced in 1981.

abrasives (3)
   see under industrial minerals
aluminum ores *
andalusite deposits* (3)
   see under ceramic materials
anhydrite deposits* (3)
   see under gypsum deposits*
antimony ores*
arsenic ores*
asbestos deposits*
asphalt (3)
   see under bitumens
barite deposits*
base metals
bauxite
bentonite deposits*
beryl (3)
   see under beryllium ores* or gems
beryllium ores*
   includes beryl (3)
bismuth ores*
bitumens
   includes asphalt (3)
borate deposits* (3)
   see under boron deposits*
boron deposits*
   includes borate deposits*
brines
   includes bromine deposits*, iodine deposits* (3)
   see also salt

```
bromine deposits*
   see also brines, salt
brown coal
   use lignite
building stone (3)
   see under construction materials, granite deposits*,
   sandstone deposits*, limestone deposits*, or
   marble deposits*
calcite deposits*
cement materials (3)
   see under limestone deposits* or construction materials
ceramic materials
   includes andalusite deposits* (3), kyanite deposits* (3),
   refractory materials (3), sillimanite deposits* (3)
chalk deposits* (3)
   see under limestone deposits*
chromite ores*
clays
   includes fuller's earth (3), and shale used as brick clay
coal
cobalt ores*
columbium ores
   use niobium ores*
construction materials
   includes perlite (3), gravel deposits*, building stone(3),
   cement materials (3), dimension stone (3)
copper ores*
corundum deposits*
cryolite deposits* (3)
   see under fluorspar
diamonds
diatomite
dimension stone (3)
   see under construction materials, granite deposits*, sand-
   stone deposits*, limestone deposits*, marble deposits*
dolostone deposits*
energy sources
   a grouper for petroleum, natural gas, coal, uranium ores*,
   etc.
evaporite deposits*
feldspar deposits*
fluorite
   use fluorspar
fluorspar
fuel resources
   a grouper for petroleum, natural gas, oil shale,
   oil sands
fuller's earth (3)
   see under clays
garnet deposits* (3)
   see under industrial minerals
gems
   includes beryl (3)
geothermal energy
glauconite deposits*
gold ores*
granite deposits*
```

graphite deposits*
gravel deposits*
   includes sands when sand is used as a construction
   material
gypsum deposits*
   includes anhydrite deposits* (3)
heavy mineral deposits*
helium gas*
hematite (3)
   see under iron ores*
industrial minerals
   includes abrasives (3), garnet deposits* (3)
iodine deposits* (3)
   see under brines, salt
iron ores*
   includes hematite (3), limonite (3), magnetite (3)
kaolin deposits*
   see also clays, bentonite deposits*
kyanite deposits* (3)
   see under ceramic materials
lead ores*
lead-zinc deposits
   see also lead ores*, zinc ores*
lignite
limestone deposits*
   includes chalk deposits* (3)
   see also dolostone deposits*
limonite (3)
   see under iron ores*
lithium ores*
magnesite deposits*
magnesium ores*
magnetite (3)
   see under iron ores*
manganese ores*
marble deposits*
mercury ores*
metal ores*
mica deposits*
mineral resources
   for very general treatments
molybdenum ores*
monazite deposits*
   see also heavy mineral deposits*, rare earth deposits*,
   thorium ores*
natural gas
nickel ores*
niobium ores*
nitrate deposits*
nonmetal deposits*
   do not use for energy sources, fuel resources*, or
   water resources
oil
   use petroleum
oil and gas fields*
   use only for detailed discussions of specific fields;
   otherwise use petroleum or natural gas

oil sands
oil shale
peat
pegmatite
perlite (3)
    see under construction materials
petroleum
phosphate deposits*
platinum ores*
polymetallic ores
potash
pumice deposits*
pyrite ores*
quartz crystal
rare earth deposits *

refractory materials (3)
    see under ceramic materials
salt
sands
    includes silica (3) for glass, ceramic, chemical use, etc.
    use gravel deposits* for sand as a construction material
sandstone deposits*
shale (3)
    use clays for shale as brick clay
    see under construction materials
silica (3)
    see sands
sillimanite deposits* (3)
    see under ceramic materials
silver ores*
slate deposits*
soapstone (3)
    see under talc deposits*
soda ash
    use sodium carbonate
sodium carbonate
    includes trona (3)
sodium sulfate
sulfur deposits*
talc deposits*
    includes soapstone (3)
tantalum ores*
tar sands
    use oil sands
thorium ores*
    see also monazite deposits*
tin ores*
titanium ores*
trona (3)
    see under sodium carbonate
tungsten ores*
uranium ores*
vanadium ores*
vermiculite deposits*
water resources
zinc ores*
zircon deposits*

D — Elements

All of the elements in the periodic table are index terms
and are valid as first-, second-, and third-level terms.
Prior to 1981, element terms were used for both economic
and geochemical papers. Starting in 1981, other terms have
been added for economic papers (see List C), and the
element terms have been restricted to geochemical papers.

Terms related to elements:
deuterium (valid on level 1, level 2, and level 3)
tritium (1,2,3)
metals (1,2,3)
nonmetals (1,2,3)
noble gases (1,2,3)
    Use for rare gases and inert gases
rare earths (1,2,3)
    Use for lanthanide series, inner transition elements,
    and lanthanoans. Includes cerium, dysprosium,
    erbium, europium, gadolinium, holmium, lanthanum,
    lutetium, neodymium, praseodymium, prometheum,
    samarium, scandium, terbium, thulium, ytterbium, and
    yttrium, all of which are (1,2,3)
major elements (3)
minor elements (3)
trace elements (3)

B.I.G. SEARCHING
    For geochemical references, see the element as a first-
    level term. Also look under the terms isotopes, chem-
    ical analysis, geochemistry, X-ray analysis, spec-
    troscopy, thermal analysis, and electron microscopy.
    Also consider using the terms related to elements, above
    Note that a cross-reference will appear in the Subject
    Index from trace elements to each first-level term under
    which it is used from 1977 on.

    For economic references, see List C, Commodities.

    _____

Sets having elements or related terms on level one:

abundance (2)
    name of material (3)
        e.g. manganese   rare earths
            abundance   abundance
                soils   granite

geochemistry (2)
    name of material (3)
        e.g. magnesium   iron
            geochemistry   geochemistry
                magmas   sedimentary rocks

analysis (2)
  type of analysis (3) such as:
  activation analysis   mass spectroscopy
  alpha-ray absorption   microwave spectroscopy
  atomic spectroscopy   Mossbauer spectroscopy
  autoradiography   neutron activation analysis
  chemical analysis   neutron diffraction analysis
  chromatography   nuclear magnetic resonance
  colorimetry   optical spectroscopy
  electrolytic analysis   polarography
  electron diffraction   radio-frequency spectroscopy
    analysis   Raman spectroscopy
  electron paramagnetic   spectroscopy
    resonance   ultraviolet spectroscopy
  electron probe   vacuum fusion analysis
  emission spectroscopy   volumetric analysis
  flame photometry   x-ray analysis
  gamma-ray spectroscopy   x-ray diffraction analysis
  ion probe   x-ray fluorescence
  laser probe   x-ray spectroscopy

    e.g. rare earths
         analysis
             neutron activation analysis

isotopes (2)
  name of isotope or isotope ratio (3)

    e.g. oxygen   lead
         isotopes   isotopes
           0-18/0-16   Pb-210

E - Geologic age (stratigraphic) terms

The authority used in GeoRef for age and stratigraphic terms
is the Geologic Time Table, compiled by F.W.B. Van Eysinga,
3rd edition, Amsterdam, Elsevier, 1975. This is followed
for nomenclature except in a few cases where previous usage
in GeoRef strongly supports alternate terms.

Throughout, the "stratigraphic" upper, middle, and lower
rather than the "chronologic" early, middle, and late are
used to modify age terms, e.g. Upper Cambrian instead of
Late Cambrian. "Uppermost" and "lowermost" are not used.

B.I.G. SEARCHING
   The terms to search in the subject index of the B.I.G.
   are those marked as possibly occurring on level one, i.e.
   those followed by (1,2,3) in the list. Under Tertiary,
   Vol. 44, No. 9, Sept. 1980, p. I-198 is the entry:

   Tertiary see also under geochronology under France;
      USSR; see also under stratigraphy under Algeria;
      Benin; California; Cuba; Czechoslovakia; England;
      France; India; Indian Ocean; Indonesia....

   Tertiary - stratigraphy
      biogeography: Tertiary 33978

   In this example, the see also reference is a guide to
   each area set in which Tertiary occurs on level three.
   Papers cited under Tertiary as a level-one term are those
   not devoted to a specific area. In the example above the
   title of the paper happens to be "Tertiary".

   Papers dealing with stratigraphy as a discipline, i.e.
   papers on its practice, philosophy, principles, etc., are
   found under the heading stratigraphy.

   Many papers on fossils contain stratigraphic information.
   If this is not a major part of the paper it may not have
   a stratigraphy set. Such papers can be found under
   appropriate fossil terms (see List F). Note that as of
   1981, a number of non-systematic terms have been added to
   index fossils in biostratigraphic papers, e.g. amphibians,
   birds, crustaceans and brachiopods. From 1981 on, these
   non-systematic fossil terms are used for stratigraphic
   papers while the systematic fossil terms are reserved for
   paleontologic papers.

   In the list which follows, the terms followed by (1,2,3)
   are valid on all levels. The remaining terms are valid
   only on level three, and are used mostly as supplemental
   index terms. Dashed lines separate parallel stratigraphic
   terminology used in different areas.

Phanerozoic (1,2,3)
  Cenozoic (1,2,3)
    Quaternary (1,2,3)
      Holocene (1,2,3)
      Flandrian
      Tyrrhenian
      Sicilian
      Calabrian
      - - - - - - - - -
      Holocene (1,2,3)
      Pleistocene (1,2,3)
        upper Pleistocene
        middle Pleistocene
        lower Pleistocene
          Villafranchian
    Tertiary (1,2,3)
      Neogene (1,2,3)
        Pliocene (1,2,3)
          upper Pliocene
          middle Pliocene
          lower Pliocene
        Miocene (1,2,3)
          upper Miocene
            Meotian
            Sarmatian
            Tortonian
          middle Miocene
            Helvetian
          lower Miocene
            Burdigalian
            Aquitanian
      Paleogene (1,2,3)
        Oligocene (1,2,3)
          Stampian
          upper Oligocene
          middle Oligocene
          lower Oligocene
        Eocene (1,2,3)
          upper Eocene
          middle Eocene
          lower Eocene
        Paleocene (1,2,3)
          Thanetian
          Montian
          Danian
  Mesozoic (1,2,3)
    Cretaceous (1,2,3)
      Upper Cretaceous
        Senonian
          Maestrichtian
          Campanian

(marine)

(continental)

Phanerozoic - Continued
  Mesozoic - Continued
    Cretaceous - Continued
      Upper Cretaceous - Continued
        Senonian - Continued
          Santonian
          Coniacian
        Turonian
        Cenomanian
      Lower Cretaceous
        Albian
        Aptian
        Neocomian
        Barremian
        Hauterivian
        Valanginian
        Berriasian
    Jurassic (1,2,3)
      Upper Jurassic
        Portlandian
          Tithonian
        Kimmeridgian
        Oxfordian

      Middle Jurassic
        Callovian
        Bathonian
        Bajocian
        Aalenian
      Lower Jurassic
        upper Liassic
          Toarcian
        middle Liassic
          Domerian
          Pliensbachian
        lower Liassic
          Sinemurian
          Hettangian
    Triassic (1,2,3)
      Rhaetian
      Upper Triassic
        Norian
        Carnian
      Middle Triassic
        Ladinian
        Anisian
      Lower Triassic
      - - - - - - - - - -
      Keuper
      Muschelkalk
      Bunter

Phanerozoic-Continued
  Paleozoic (1,2,3)
    Permian (1,2,3)
      Thuringian
      Saxonian
      Autunian
      - - - - - - - - -
      Zechstein
      Rotliegendes
      - - - - - - - - -
      Tatarian
      Kazanian
      Kungurian
      Artinskian
      Sakmarian
      - - - - - - - - -
      Upper Permian
      Lower Permian
    Carboniferous (1,2,3)
      Upper Carboniferous
        Stephanian
        Westphalian
        Namurian
        Dinantian
        Visean
        Tournaisian
      - - - - - - - - -
      Upper Carboniferous
        Gzhelian
        Kasimovian
        Moscovian
        Bashkirian
        Namurian
        Dinantian
        Visean
        Tournaisian
      - - - - - - - - -
    Pennsylvanian (1,2,3)
      Upper Pennsylvanian
      Middle Pennsylvanian
      Lower Pennsylvanian
    Mississippian (1,2,3)
      Upper Mississippian
      Middle Mississippian
      Lower Mississippian

Phanerozoic-Continued
  Paleozoic-Continued
    Devonian (1,2,3)
      Upper Devonian
        Strunian
        Famennian
        Frasnian
      Middle Devonian
        Givetian
        Eifelian
      Lower Devonian
        Emsian
        Siegenian
        Gedinnian
    Silurian (1,2,3)
      Upper Silurian
        Ludlovian
        Wenlockian
      Lower Silurian
        Tarannon
        Llandoverian
    Ordovician (1,2,3)
      Upper Ordovician
        Ashgillian
        Caradocian
      Middle Ordovician
        Llandeilian
        Llanvirnian
      Lower Ordovician
        Arenigian
        Tremadocian
    Cambrian (1,2,3)
      Upper Cambrian
      Middle Cambrian
      Lower Cambrian

Precambrian (1,2,3)
  Proterozoic (1,2,3)
    upper Proterozoic
      Vendian
    middle Proterozoic
    lower Proterozoic
  Archean (1,2,3)

F – Fossils

The terminology in this List was developed in consultation with micropaleontologists at the American Museum of Natural History for whom we produce the monthly Bibliography and Index of Micropaleontology, and with vertebrate paleontologists at the Museum of Paleontology of the University of California-Berkeley with whom we produce the annual Bibliography of Fossil Vertebrates.

All terms in the List are systematic names except those on the right margin under the heading 'Common Names'. The latter were reserved in 1981 for biostratigraphy papers.

INDEXING
Terms on the left margin are used on all levels. Terms indented from the left are used on levels two or three only, except Graptolithina and Pterobranchia, which are also valid on level one. Terms on the right margin (common names) which are opposite terms on the far left margin are valid on levels one and three. The remaining common names are valid on level three only. Use the most specific term applicable.

Several codes are provided. The numbers from 8-11 stand for the section in the Bibliography and Index of Geology for that fossil group:

08 - general paleontology
09 - paleobotany
10 - invertebrate paleontology
11 - vertebrate paleontology

The letter 'V' under a term indicates that a citation with that index term is to be given a 'V' Special Bibliography Code to mark it for the Bibliography of Fossil Vertebrates.

A three character 'M' code under a term indicates that the citation of a paper in which that fossil group is the main topic should have that Special Bibliography Code for the Bibliography and Index of Micropaleontology sections:

M01	algae	M07	problematic
M02	scolecodonts		microfossils
M03	nannofossils;	M08	Ostracoda
	Coccolithophoraceae	M09	palynomorphs
M04	Conodonta	M10	Protista
M05	diatoms	M11	Radiolaria
M06	foraminifera	M12	miscellanea

Fossil sets are constructed as follows:

fossil group (1)
    fossil subgroup (2)
        age (List E) (3)

e.g. Bryozoa   foraminifera
        Cheilostomata   Globigerinacea
        Miocene   Cretaceous

If the age is not known, an area term should be used:

fossil group (1)
  fossil subgroup (2)
    area (List 0) (3)

e.g. Mollusca
      Cephalopoda
      France

If neither age nor area are known, use a topic, e.g.
morphology, on level three. For a paper about more than
one fossil subgroup, e.g. Buliminacea and Carterinacea,
use an available broader subgroup on level two, e.g.
Rotaliina.

For papers on more than one major subgroup, e.g. Rotaliina
and Miliolina, construct the set as follows:

fossil group (1)

  topic (2)
  (biocenoses, biochemistry, biogeography, biostratigraphy (see below), distribution, ecology, evolution
  (including phylogeny), faunal studies, fossilization
  (including taphonomy), habitat, miscellanea,
  morphology, paleoecology, occurrence, ontogeny,
  taxonomy.

  topic (3)
  age (List E), benthonic taxa, bibliography,
  biometry, catalogs, faunal list, floral list,
  paleoclimatology, phylogeny, planktonic taxa,
  statistical analysis, taphonomy, thanatocenoses,
  zoning, or type of material, e.g. bones, fossil
  wood, plankton, shells, skeletons, skulls, teeth,
  or tests.

e.g. Mammalia   Reptilia   Ostracoda
      faunal studies taxonomy   paleoecology
      Cenozoic   morphology   paleoclimatology

Beginning in 1981, terms on the right margin, common
names, are used on level one instead of their systematic
level-one equivalents for biostratigraphy papers.

e.g. ferns
      biostratigraphy
      Cretaceous

If there is no common name term on the List, use the
equivalent systematic term in the set and enter a
proposed common name as a supplemental index term.

e.g. Arthropoda
      biostratigraphy
      Cambrian

      arthropods (supplemental index term)

For paleoecology and biogeography sets having fossil
terms on level two, use systematic names only. These
sets are not to be used for stratigraphy papers.

e.g.   paleoecology   biogeography
       Radiolaria   Pisces
       Paleogene   Mesozoic

For papers which emphasize both paleontology and biostratigraphy, two fossil sets may be used.

e.g.   Pisces   fish
       morphology   AND   biostratigraphy
       taxonomy   Tertiary

B.I.G. SEARCHING

Both the systematic and, since 1981, the common names of
fossil groups are entry terms in the subject index. The
sets with these terms on level one are subdivided on
level two by either narrower fossil terms or topics (see
under Indexing, above).

Related headings include paleontology (for general treatments, discussions of evolution, etc.), biogeography,
paleoecology, paleoclimatology, paleobotany, (the
discipline), palynology (subdiscipline or technique),
micropaleontology (subdiscipline or technique).

Fossil Group	Common Name
Agnatha	
11, V	
Algae	algal flora
09,(M01),	
(M03),(M05)	
Chlorophyta	
Charophyta	
Chlorophyceae	
Codiaceae	
Dasycladaceae	
Receptaculitaceae	
Desmidiales	desmids
Tasmanites	
Chrysophyta	
M03 Coccolithophoraceae	coccoliths
Cyanophyta	
M05 diatoms	diatom flora
M03 nannofossils (incl.	nannoconids
discoasters, Nannoconus)	
Phaeophyta	
Pyrrhophyta	
Rhodophyta	
Corallinaceae	
Gymnocodiaceae	
stromatolites	
Amphibia	amphibians
Labyrinthodontia (incl.	
11, V Anthracosauria,	
Ichthyostegalia, Temnospondyli)	
Lepospondyli (incl. Aistropoda,	
Microsauria, Nectridea)	
Lissamphibia (incl. Anura, Apoda,	
Gymnophiona, Caecilia,	
Proanura, Urodela)	

angiosperms		angiosperm flora
09	Dicotyledoneae	
	Asteridae	
	Caryophyllidae	
	Dilleniidae	
	Hamamelididae	
	Magnoliidae	
	Rosidae	
	Monocotyledoneae	
	Alismidae	
	Arecidae	
	Commelinidae	
	Liliidae	
Archaeocyatha		
10		
Arthropoda		
10	Chelicerata	
	Arachnida	
	Merostomata	
	Mandibulata	
	Crustacea	crustaceans
	Branchiopoda	
	Cirripedia	
	Copepoda	
	Malacostraca	
	Myriapoda	
	Trilobitomorpha	
Aves		birds
	Archaeornithes	
11,V	Neornithes	
bacteria		
09		
Brachiopoda		brachiopods
10		
	Articulata	
	Orthida	
	Dictyonellidina	
	Pentamerida	
	Rhynchonellida	
	Spiriferida	
	Strophomenida	
	Terebratulida	
	Thecideidina	
	Inarticulata	
bryophytes		
09	Hepaticae	
	Musci	
Bryozoa		bryozoans
10		
	Cheilostomata	
	Cryptostomata	
	Ctenostomata	
	Cyclostomata	
	Trepostomata	
Chordata	(incl. Protochordata, Vertebrata)	
11, V		

```
Coelenterata                                          corals
10              Anthozoa
                  Actiniaria
                  Ceriantipatharia
                  Corallimorpharia
                  Heterocorallia
                  Hexactiniaria
                  Octocorallia
                  Rugosa
                  Scleractinia
                  Tabulata
                  Zoanthiniaria
                Hydrozoa
                Scyphozoa
                  Conularida
                Stromatoporoidea
Conodonta                                             conodonts
08,M04
coprolites
08,11,(V)
Echinodermata                                         echinoderms
10              Asteroidea
                Blastoidea
                Camptostromatoidea
                Crinoidea
                Cyclocystoidea
                Cystoidea
                Echinoidea
                Edrioasteroidea
                Edrioblastoidea
                Eocrinoidea
                Helicoplacoidea
                Holothuroidea
                Homoiostelea
                Homostelea
                Lepidocystoidea
                Machaeridia
                Ophiocistioidea
                Ophiuroidea
                Parablastoidea
                Paracrinoidea
                Somasteroidea
                Stylophora
foraminifera                                          foraminifers
10, M06         Allogromiina
                Fusulinina
                  Fusulinidae
                Miliolina
                  Alveolinellidae
                  Miliolacea
                Rotaliina
                  Buliminacea
                  Carterinacea
                  Cassidulinacea
                  Discorbacea
                  Globigerinacea
                  Nodosariacea
```

```
foraminifera-continued
                Rotaliina-Continued
                Orbitoidacea
                  Orbitoididae
                Robertinacea
                Rotaliacea
                  Nummulitidae
                Spirillinacea
                Textulariina
                Ammodiscacea
                Lituolacea
fossil man                Use only when fossil remains are
11, V                     discussed; otherwise use archaeology
                          or anthropology under stratigraphy
                          (2nd level) under area terms.

fungi
09              Myxomycetes

gymnosperms                                    gymnosperm flora
09              Bennettitales
                Caytoniales
                Coniferales
                Cordaitales
                Cycadales
                Cycadofilicales
                Ephedrales
                Ginkgoales
                Glossopteridales
                Gnetales
                Nilssoniales
                Pentoxylales
                Welwitschiales
Hemichordata Enteropneusta
10              Graptolithina   (1)                graptolites
                  Camaroidea
                  Dendroidea
                  Graptoloidea
                    Didymograptina
                    Diplograptina
                    Glossograptina
                    Monograptina
                  Stolonoidea
                  Tuboidea
                Pterobranchia (1)
                  Cephalodiscida
                  Rhapdopleurida
ichnofossils
08,09,10,11,          (Use lebensspuren for sedimentary
(V)                       structures)
Insecta         Blattopteroida      (Blattodea)
10              Coleopteroida       (Coleoptera)
                Dermapteroida       (Dermaptera)
                Ectotropha
                Entotropha
                Ephemeropteroida    (Ephemeroptera)
                Hemipteroida        (Hemiptera)
                Hymenopteroida      (Hymenoptera)
                Lepidopteroida      (Lepidoptera)
```

Insecta-Continued
```
        Mecopteroida           (Mecoptera)
        Neuropteroida          (Neuroptera)
        Odonatopteroida        (Odonata)
        Orthopteroida          (Orthoptera)
        Palaeodictyopteroida   (Palaeodictyoptera)
        Psocopteroida          (Psocoptera)
        Siphonapteroida        (Siphonaptera)
        Thysanopteroida        (Thysanoptera)
Invertebrata                                        invertebrates
10
lichens
09
Mammalia                                            mammals
11, V   Acreodi (incl. Mesonychoidea)
        Anagalida
        Arctocyonia
        Artiodactyla
          Ruminantia
          Suiformes
          Tylopoda
        Astrapotheria
        Carnivora
          Fissipeda
          Pinnipedia
        Cetacea
        Chiroptera
        Condylarthra
        Creodonta
        Dermoptera
        Desmostylia
        Dinocerata
        Diprotodonta
        Docodonta
        Edentata
        Embrithopoda
        Hyracoidea
        Insectivora
        Lagomorpha
        Litopterna
        Macroscelida
        Marsupialia
        Monotremata
        Multituberculata
        Nomarthra
        Notoungulata
        Pantodonta
        Paucituberculata
        Pantotheria (incl. Eupantotheria)
        Perissodactyla
          Ceratomorpha
          Hippomorpha
        Pholidota
        Polyprotodontia
```

Mammalia-Continued
        Primates
          Hominidae
            Homo sapiens
            Neanderthal
            pre-Neanderthal
          Pongidae (incl. Apes)
          Prosimii (incl. Lemuroidea,
                  Tarsioidea,
                  Pleisiadapoidea,
                  Anthropoidea)
          simians
            Cynomorpha
            Platyrrhina
        Proboscidea
        Barytherioidea
        Deinotherioidea
        Elephantoidea
        Mastodontoidea
        Moeritherioidea
        Proteutheria
        Pyrotheria
        Rodentia   rodents
          Hystricomorpha
          Myomorpha
          Sciuromorpha
        Scandentia
        Sirenia
        Symmetrodonta
        Taeniodonta
        Theria
        Tillodontia
        Triconodonta
        Tubulidentata
        Xenarthra
        Xenungulata
Mollusca   mollusks
10
        Aplacophora
        Bivalvia   bivalves
          Actinodontida
          Astartida
          Carditida
          Ctenodontida
          Cyrtodontida
            Arcina
            Ostreacea
            Pteriina
              Inocerami
          Pectinacea
          Pholadomyida
          Praecardiida
          Rudistae   rudists
          Septibranchia
          Solemyida
          Venerida

Mollusca-Continued
        Cephalopoda   cephalopods
          Tetrabranchiata
            Ammonoidea   ammonoids
              Anarcestida
              Bactritida
              Ceratitida
              Clymeniida
              Desmoceratida
              Goniatitida
              Lytoceratida
              Perisphinctida
              Phylloceratida
              Prolecanitida
              Psiloceratida
            Nautiloidea
          Dibranchiata
            Belemnoidea
        Gastropoda   gastropods
         Acoela
         Archaeogastropoda
         Basommatophora
         Bellerophontina
         Entomotaeniata
         Mesogastropoda
         Neogastropoda
         Pteropoda
         Sacoglossa
         Stylommatophora
        Hyolithes
        Monoplacophora
        Polyplacophora
        Rostroconchia
        Scaphopoda
        Tentaculites
Ostracoda   ostracods
10,M08   Archeocopida
        Bairdiomorpha
        Beyrichicopina
        Cladocopina
        Cypridocopina
        Cytherocopina
        Entomozocopina
        Eridostraca
        Kirkbyocopina
        Leperditicopida
        Myodocopina
        Platycopida
palynomorphs   palynomorphs
09,M09   acritarchs   acritarch flora
        Chitinozoa
        Dinoflagellata   dinoflagellates
        megaspores
        miospores

Pisces		fish
11, V	Chondrichthyes	
	Elasmobranchii	
	Eubradyodonti	
	Holocephali	
	Osteichthyes	
	Actinistia	
	Brachiopterygii	
	Chondrostei	
	Dipnoi	
	Halecostomi	
	Holostei	
	Rhipidistia	
	Teleostei	
	Placodermi	
Plantae		
09		
Porifera		
10	Calcispongea	
	Demospongea	
	Hyalospongea	
problematic fossils		
08,10,11,V, M07		
	problematic microfossils	
Protista		
08,09,10,M10	ebridians	
	Silicoflagellata	
	Thecamoeba	
	Tintinnidae	
pteridophytes		ferns
09	Filicopsida	
	Lycopsida	
	Noeggerathiales	
	Psilopsida	
	Pteridophyllen	
	Sphenopsida	
Radiolaria		radiolarians
10,M11	Acantharina	
	Nassellina	
	Phaeodarina	
	Spumellina	
Reptilia		reptiles
11, V	Anapsida	
	Captorhinomorpha	
	Chelonia	
	Cotylosauria	
	Pareiasauria	
	Archosauria	
	Crocodilia	
	Ornithischia	
	Pterosauria	
	Saurischia	
	Thecodontia	
	dinosaurs	

```
            Reptilia-Continued
                        Euryapsida
                         Araeoscelidia
                          Trilophosauria
                         Sauropterygia
                          Nothosauria
                          Plesiosauria
                         Ichthyopterygia
                          Ichthyosauria
                         Lepidosauria
                          Eosuchia
                          Rhynchocephalia
                          Squamata
                         Synapsida
                          Mesosauria
                           Proganosauria
                          Pelycosauria
                          Placodontia
                          Therapsida
            Spermatophyta  (incl. angiosperms and gymnosperms)
            09
            Tetrapoda      (incl. Aves, Amphibia, Reptilia, Mammalia)
            11, V
            thallophytes   (incl. bacteria, algae, fungi, lichens)
            09
            Trilobita                                  trilobites
            10             Agnostida
                           Corynexochida
                           Lichida
                           Odontopleurida
                           Phacopida
                           Ptychopariida
                           Redlichiida
            Vertebrata                                 vertebrates
            11, V
            worms
            10             Annelida
                           Chaetognatha
                           Echiurida
                           Myzostomia
                           Nematoida
                           Nematomorpha
                           Nemerta
                           Oligochaetia
                           Phoronida
                           Pogonophora
                           Polychaetia
                           Priapulida
            M02            scolecodonts
                           Sipunculoida
            - - - - - - - - - - - - - - - - - - - - - - - - - - - -
                           pollen analysis (results)
                           palynology      (technique)
```

G - General terms

INDEXING
The following general terms are to be used sparingly.
They are permitted in certain sets on levels two and
three.

B.I.G. SEARCHING
Look for these terms on the second level under topics
of interest.

---

age
annual report
anomalies
applications
bibliography
catalogs
causes
changes
classification
composition
concepts
detection
distribution
education
effects
environment
evolution
experimental studies
general
genesis
history
identification
instruments
interpretation
manuals
mechanism

methods
nomenclature
objectives
observations
occurrence
origin, use genesis
patterns
philosophy
practice
preparation
principles
processes
properties
rates
regional
review
surveys (different from
    survey organizations)
symposia
techniques
temperature
textbooks
textures
theoretical studies
varieties

H – Igneous rocks

INDEXING

The following list indicates the second-level terms permitted in sets under the term igneous rocks. These terms are also permitted on the second level under phase equilibria.

The most specific term is to be used on the second level. Should the paper under consideration deal with several rocks belonging to different groups, a general topic should be selected (see igneous rocks in the Term List).

Singular forms of terms such as granite may be used as index terms but not in igneous rocks sets.

e.g. igneous rocks		igneous rocks
granites	not	granites
textures		granite

Should there be no need for an igneous rocks set (i.e. if these rocks are present but not emphasized in the paper) use specific rock names in the singular form,

e.g. alaskite, latite, microdiorite, etc.

B.I.G. SEARCHING

At the entry for igneous rocks, look under the appropriate second-level term(s) selected from the list below.

Look under related headings such as intrusions, isotopes, magmas, metamorphic rocks, metamorphism, metasomatic rocks, metasomatism, and phase equilibria.

Look under petrology for papers dealing with the discipline, such as papers on techniques, practice of, etc.

igneous rocks
  plutonic rocks (2)
  hypabyssal rocks (2)

  granites (2)
    microgranite
    alkali granites (2)
    granodiorites (2)
    charnockite

      aplite
  diorites (2)
    microdiorite
    alkali diorites (2)
    quartz diorites (2)
  syenites (2)
    microsyenite
    alkali syenites (2)
    monzonites (2)
  gabbros (2)
    microgabbro
    alkali gabbros (2)
      nephelinite
      leucitite
    anorthosite
    norite
  ultramafics (2)
    pyroxenite
    peridotites (2)
    kimberlite
  lamprophyres (2)
  carbonatites (2)
  diabase (2)

  volcanic rocks (2)
    rhyolites (2)
    rhyodacites (2)
    dacites (2)
    andesites (2)
    trachyandesites (2)
    trachytes (2)
      keratophyre
    phonolites (2)
    basalts (2)
      alkali basalts (2)
        trachybasalts (2)
    pyroclastics (2)
      ignimbrite
      perlite
      pumice
      tuff
      tuffite
    glasses (2)
      obsidian

I - Sedimentary rocks

INDEXING

The term <u>sedimentary rocks</u> is used as a first-level
term in a set. Terms from the following list are used
on the second and third levels.

e.g. sedimentary rocks
        clastic rocks
          sandstone

The most specific second-level term is used except
when the paper discusses several rock groups, or general
topics such as <u>textures</u>, in which case the second-level
term should be a topic:

e.g. sedimentary rocks
        textures
          grain size

Other related sets include <u>sediments</u>, <u>sedimentation</u>,
and <u>sedimentary</u> structures.

If the rocks are not emphasized, a set for sedimentary
rocks is not needed, but the rock terms or rock group
should be used as index terms.

B.I.G. SEARCHING

At the entry for <u>sedimentary rocks</u>, look under the
appropriate second-level term(s) selected from the list
below.

Look under related headings such as <u>sedimentation</u>,
<u>sedimentary</u> structures, <u>sediments</u>, <u>diagenesis</u>, and
<u>weathering</u>.

For papers dealing with the discipline, such as papers
on techniques, practice of, etc., look under <u>sedimentary</u>
<u>petrology</u>.

Look under specific sedimentary rocks which are
commodities, e.g. <u>limestone</u> <u>deposits</u> (see List C,
Commodities).

---

sedimentary rocks
  carbonate rocks (2)
    algal limestone
    beachrock
    biocalcarenite
    calcarenite
    calcrete
    chalk
    coquina
    dolomitic limestone
    dolostone
    limestone
    micrite
    oolitic limestone
    travertine

sedimentary rocks-continued
  chemically precipitated rocks (2)
    chert
      jasperoid
    evaporites
      anhydrite
      gypsum
      potash
      salt
      ferricrete
      flint
      iron formations
      phosphate rocks
        bone beds
      silcrete
      siliceous sinter
  clastic rocks (2)
    argillite
    arkose
    bauxite
    bentonite
    breccia
        tectonic breccia
        volcanic breccia
    conglomerate
    diatomaceous earth
    flysch
    graywacke
    marl
    molasse
    radiolarite
    red beds
    sandstone
    shale
    siltstone
    spongolite
    tillite
    tonstein
    turbidite
  organic residues (2)
    coal
        anthracite
        bituminous coal
        coke coal
        lignite
        sapropelite
        steam coal
      macerals
        durain
        exinite
        inertinite
        vitrain
        vitrinite
      oil sands
      oil shale

J – Metamorphic rocks

INDEXING

Metamorphic rocks is a first-level term with terms from the follow
list used on the second level. These terms can also be used on th
second level under the first-level term phase equilibria.

Make a metamorphic rocks set only when the rocks are emphasized.
Otherwise, add metamorphic rocks or a second-level term from the l
below as a supplemental index term, without constructing a set.

For documents on petrology as a discipline, make a petrology set.

Make a metamorphism set for papers which stress the process.

A specific rock group should be used on the second level unless th
paper discusses several groups and/or general topics such as miner
assemblages. For such papers, the second level should be a topic:

```
e.g.  metamorphic rocks
          mineral assemblages
              hornblende schist
```

No singular form of the group names is allowed as in index term.
Use amphibolites not amphibolite, and gneisses not gneiss.

```
e.g.  metamorphic rocks
          gneisses
              chemical composition
```

B.I.G. SEARCHING

At the entry for metamorphic rocks, look under the appropriate
second-level term(s) selected from the list below.

Look under related headings such as metamorphism, petrology,
metasomatism, phase equilibria, and paragenesis.

For papers dealing with the study of metamorphic rocks,
look under petrology.

---

metamorphic rocks
  metasedimentary rocks
  metavolcanic rocks
  metaplutonic rocks
  metaigneous rocks
  hornfels
  slates
  phyllites
  schists
   greenschist
   blueschist
  gneisses
   orthogneiss
   paragneiss

amphibolites
  orthoamphibolite
  para-amphibolite
granulites
marbles
mylonites
cataclasites
phyllonites
migmatites
  embrechite
  anatexite
quartzites
garnetite
eclogite
itabirite

K – Sedimentary structures

INDEXING
Use specific terms from the following list. If a set
needs to be constructed, use the general structure type
on level two and the specific name on level three.

B.I.G. SEARCHING
At the entry for sedimentary structures, look under the
appropriate level-two and level-three term(s) selected
from the following list.

Also under sedimentary structures, look for topics on
level two, such as environmental analysis.

———————————

sedimentary structures
  bedding plane irregularities
    antidunes
    chevron marks
    (current lineations use parting lineation)
    current markings
    (current partings use parting lineation)
    flute casts
    frost features
    grooves
    groove casts
    ice wedges
    load casts (also under soft sediment
      deformation and turbidity current structures)
    megaripples
    mounds
    mudcracks
    mud lumps
    parting lineation
    ripple marks (also under primary structures)
      (asymmetrical, interference, etc.)
    sand waves
    scour casts
    scour marks
    sole marks (also under turbidity current
      structures)
    striations
    tool marks
  biogenic structures
    algal banks
    algal biscuits
    algal mats
    algal mounds
    algal structures
    banks
    bioturbation
    burrows
    coprolites
    girvanella

        sedimentary structures - continued
         biogenic structures - continued
          lebensspuren
          oncolites
          stromatolites
          tracks
          trails
         cylindrical structures
         planar bedding structures (i.e. bedding
           structures)
          bars
          bedding (also under primary structures)
          channels
          cross-bedding
          cross-laminations
          cross-stratification
          cut and fill
          cyclothems
          flaser bedding
          graded bedding (also under turbidity
            current structures)
          imbrication
          laminations
          massive bedding
          megacyclothems
          rhythmic bedding
          ripple drift-cross laminations
          (ripple-cross-laminations use ripple
            drift-cross laminations)
          sand bodies
          stratification
          varves
         primary structures
          bedding (also under planar bedding structures)
          ripple marks (also under bedding plane
            irregularities) (asymmetrical, interference,
            etc.)
         secondary structures
          armored mud balls
          concretions
          cone-in-cone
          geodes
          microstylolites
          septaria
          stylolites
         soft sediment deformation
          ball-and-pillow
          boudinage
          clastic dikes
          convoluted beds (also under turbidity current
            structures)
          (convoluted deformation use convoluted beds)
          flame structures
          flow structures (also under turbidity current
            structures)

sedimentary structures - continued
  soft sediment deformation - continued
    load casts (also under bedding plane
      irregularities and turbidity current
      structures)
    olistoliths (also under turbidity current
      structures)
    olistostromes (also under turbidity current
      structures)
    (pull apart structures use boudinage)
    sandstone dikes
    slump structures
  turbidity current structures
    convoluted beds (also under soft sediment
      deformation)
    (convoluted deformation use convoluted beds)
    flow structures (also under soft sediment
      deformation)
    graded bedding (also under planar bedding
      structures)
    load casts (also under bedding plane
      irregularities and soft sediment
      deformation)
    olistoliths (also under soft sediment
      deformation)
    olistostromes (also under soft sediment
      deformation)
    sole marks (also under bedding plane
      irregularities)

L – Minerals

INDEXING

The following group names (chemical for the non-silicates
and structural for the silicates) are used on the second
level under minerals, crystal chemistry, crystal
structure, phase equilibria, and crystal growth. On the
third level a specific mineral name (if the paper deals
with one mineral primarily) or a topic such as composition
or crystal chemistry (if the paper deals with more than
one mineral) is used.

A mineral set is required whenever the paper deals with
mineralogic aspects of a rock or with a specific mineral
in a substantial manner. Papers discussing petrology
with simple mention of mineral composition should not
receive a mineral set. Rather, the mineral names
involved in that case are used simply as index terms.
Thus, a paper about the crystal structure of quartz will
definitely have the following two sets:

minerals
    framework silicates, silica minerals
        quartz

crystal structure
    framework silicates, silica minerals
        quartz

But a paper discussing quartz distribution in a sandstone
will simply have quartz as an index term.

In cases where the particular mineral species may be
referred to more than one group (e.g. kainite belonging
to both halides and sulfates), two minerals sets are made.

Papers dealing with the disciplines of mineralogy and
crystallography are indexed using those terms on the 1st
level.

e.g. mineralogy   crystallography
         practice   research
            mineralogists   current research

For papers dealing with new mineral species, the term
new minerals is used as an index term, and the chemical
formula is given as an index term if available.

B.I.G. SEARCHING

The headings minerals, mineralogy, crystal structure,
crystal chemistry, crystal growth, phase equilibria, clay
mineralogy, and crystallography are the main access points
for mineralogical papers. Other related headings are
igneous rocks, metamorphic rocks, sedimentary rocks,
metasomatic rocks, sediments, geochemistry, and paragenesis.

For major mineral groups, look on the second level under the
headings minerals, crystal structure, crystal chemistry,
crystal growth, and phase equilibria. Also cross-references
are provided to direct your attention to these groups.

e.g. sulphur   sulfosalts
        see sulfur   see under minerals

For the economic geology of minerals, look under the
appropriate minerals selected from List C, Commodities.

---

minerals
  alloys (including carbides, nitrides,
    phosphides, silicides)
  antimonates
  antimonides
  antimonites
  arsenates
  arsenides
  arsenites
  bismuthides
  borates
  bromides
  carbides
  carbonates
  chlorides
  chromates
  fluoborates
  fluorides
  fluosilicates
  halides (includes bromides, chlorides,
    fluoborates, fluosilicates, iodides, and
    fluorides)
  iodates
  iodides
  miscellaneous minerals (used for several
    groups or for minerals of unknown affinity)
  molybdates
  native elements
  niobates
  niobotantalates
  nitrates
  nitrides
  organic compounds
  oxalates
  oxides (including niobates, niobotantalates,
    tantalates)
  oxysulfides
  phosphates
  phosphides
  selenates
  selenides
  selenites
  silicates (use a narrower term below if
    dealing with specific mineral; otherwise
    larger group)
    aluminosilicates
    orthosilicates
      sorosilicates
        orthosilicates, epidote group
        orthosilicates, melilite group

minerals-continued
  silicates-continued
    orthosilicates-continued
      nesosilicates
        orthosilicates, garnet group
        orthosilicates, humite group
        orthosilicates, olivine group
      ring silicates
      chain silicates (inosilicates)
        chain silicates, amphibole group
          chain silicates, alkalic amphibole
          chain silicates, clinoamphibole
          chain silicates, orthoamphibole
        chain silicates, pyroxene group
          chain silicates, alkalic pyroxene
          chain silicates, clinopyroxene
          chain silicates, orthopyroxene
      sheet silicates (phyllosilicates)
        sheet silicates, chlorite group
        sheet silicates, clay minerals
        sheet silicates, mica group
        sheet silicates, serpentine group
      framework silicates (tectosilicates)
        framework silicates, feldspar group
          framework silicates, alkali feldspar
          framework silicates, plagioclase
          framework silicates, barium feldspar
        framework silicates, nepheline group
        framework silicates, scapolite group
        framework silicates, silica minerals
        framework silicates, sodalite group
        framework silicates, zeolite group
  silicides
  sulfates
  sulfides (including antimonides, arsenides,
    bismuthides, oxysulfides, selenides, and
    tellurides)
  sulfosalts (including sulfantimonates,
    sulfantimonites, sulfarsenates, sulfarsenites,
    sulfobismuthites, sulfogermanates, sulfo-
    stannates, and sulfovanadates)
  tantalates
  tellurates
  tellurides
  tellurites
  tungstates
  vanadates

NOTES
  1. Papers on artificial minerals are usually not treated
      except when a mineralogic application is clearly
      indicated. The indexing term synthetic materials is
      then added.

  2. Papers about mineral collecting will have the term
      collecting in the indexing.

M — Soils

INDEXING
The classification of soils has undergone several changes
in American practice. What follows are two of the
classifications using different criteria. The 1965
classification has been retained fairly intact since that
time. In addition there follows a comparative table
giving equivalent French and German terms for some of the
basic soil terms in English. We do not attempt to
standardize the soils. Use the soil terms given in the
source documents, without attempting to interpret types
of soil according to one of the available classifications.

The terms in the list below are used on level three under
soils (1) and soil group (2). In addition, terms such
as loam, laterites, Paleosols, and volcanic ash may be
used on level three.

B.I.G. SEARCHING
All the soil entries may be found under the term soils.
On the second level the term surveys followed by an area
will occur, if the article discusses aspects of soils in
a specific region. Or soil group will occur on the
second level with one of the terms from the list below
on the third level. Otherwise, a topic such as geochemistry or water regimes will be found on the second level.

AMERICAN SOIL CLASSIFICATION SYSTEM - 1964

Zonal soils
  Tundra soils
  Subarctic brown forest soils
  Desert soils
  Red desert soils
  Sierozems
  Brown soils
  Reddish brown soils
  Chestnut soils
  Reddish chestnut soils
  Chernozems
  Brunizems
  Reddish prairie soils
  Noncalcic brown soils
  Podzols
  Brown podzolic soils
  Gray wooded soils
  Sols bruns acides
  Gray-brown podzolic soils
  Red-yellow podzolic soils
  Reddish-brown lateritic soils
  Yellowish-brown lateritic soils
  Latosols

American soil classification-1964-Continued

Intrazonal soils
  Solonchak soils (saline soils)
  Solonetz soils
  Soloth soils
  Humic-gley soils
  Alpine meadow soils
  Bog soils
  Low-humic gley soils
  Planosols
  Ground-water podzols
  Ground-water laterite soils
  Brown forest soils
  Rendzinas
  Grumusols
  Calcisols

Azonal soils
  Lithosols
  Regosols
  Alluvial soils

AMERICAN SOIL CLASSIFICATION SYSTEM - 1965 et seq.

Alfisols   Aquolls
  Udalfs   Borolls
  Boralfs   Redolls
  Ustalfs   Udolls
  Xeralfs   Ustolls
                                  Xerolls
Aridisols
  Argids   Oxisols
  Orthids   Aquox
                                  Humox
Entisols   Orthox
  Aquents   Torrox
  Fluvents   Ustox
  Orthents
  Psamments   Spodosols
                                  Aquods
Histosols   Humods
  Fibrists   Orthods
  Folists
  Hemists   Ultisols
  Saprists   Aquults
                                  Humults
Inceptisols   Udults
  Andepts   Ustults
  Aquepts   Xerults
  Ochrepts
  Tropepts   Vertisols
  Umbrepts   Torrerts
                                  Uderts
Mollisols   Usterts
  Albolls   Xererts

AMERICAN	GERMAN	FRENCH
Acrisols	Acrisol	Sol-mediterraneen
Albolls		
Alfisols		
Alluvial soils	Auen-Boden	Sol-d'alluvions
Alpine meadow soils	Alpiner Wiesenboden	Sol-hydromorphe
Andepts		
Andosols	Andosol	Sol-peu-evolue roche-volcanique
Aqualfs		
Aquents		
Aquepts		
Aquods		
Aquolls		
Aquox		
Aquults		
Arctic tundra soils	Arktische Tundra Boden	Sol-de-toundra
Arenosols	Arenosol	Sol-brut sable
Arents		
Argids		
Aridisols		
Azonal soils	Roh-Boden	Sol-brut
Black earth use Chernozems	Schwarzerde	Chernozem
Bog soils	Moorboden	Tourbe
Boralfs		
Boreal frozen taiga soils		Sol-gele
Boreal taiga and forest soils		Sol
Borolls		Mollisol
Brown desert steppe soils	Burozem	Sierozem
Brown forest soils	Brauner Wald Boden	Sol-brun
Brown loam	Braunlehm	Couche-rouge
Brown podzolic soils	Podsoliger Braunerde	Podzol
Brown soils	Braunerde	Sol-brun
Brunizems	Brunizem	Brunizem
Calcisols		
Cambisols	Cambisol	Sol-brun
Caar	Uebergangsmoor	Tourbe
Chernozems	Chernozem	Chernozem
Chestnut soils	Kastannozem	Sol-chatain
Cryosols	Cryosol	Sol-gele
Desert soils	Wuesten-Boden	Sol-subdesertique
Desert raw soils	Wuesten-roh-Boden	Sol-de-desert
Dy	Dy	Vase matiere-organique
Entisols		
Fen		
Ferralites	Eisen-Silikat-Boden	Sol-fersialitique
Ferralsols	Ferralsol	Sol-ferralitique
Ferrods		
Ferruginous soils	Eisenhaltiger-Boden	Sol-ferrugineux
Fibrists		
Fluvents		

AMERICAN	GERMAN	FRENCH
Fluviosols		Sol-d'alluvions
Folists		
Gleys	Gley	Gley
Glensols	Glensol	Glensol
Gray-brown podzolic soils	Fahlerde	Sol-brun lessivage
Gray podzolic soils	Podsolierter Grauer Boden	Podzol
Gray warp soils	Paternia	Sol-peu-evolue or sol-d'alluvions
Gray wooded soils		
Ground-water podzols	Gley-Podsol	Podzol
Ground-water laterite soils	Grundwasser-Laterite	Laterite
Grumosols	Grumosol	Vertisol
Half bog soils	Anmoor	Tourbe
Halomorphic soils	Salz-Boden	Sol-halomorphe
Halosols	Halosol	Sol-halomorphe
Hemists		
High moor	Hochmoor	Tourbe
Histosols		
Humic gley soils	Humus Gley Boden	Sol-hydromorphe
Humic soils	Humus-reiche-Boden	Sol-riche-en-humus
Humods		
Humox		
Hydromorphic soils	Hydromorpher-Boden	Sol-riche-en-humus
Inceptisols		
Intrazonal soils	Intrazonaler Boden	Sol
Kastanozems		
Krasnozems	Krasnozem	Krasnozem
laterites	Laterit-Boden	Sol-ferralitique
Latosols	Latosol	Sol-ferralitique
Lithosols	Gesteins-roh-Boden	Sol-squelettique
Low-humic gley soils		
Luvisols	Luvisols	Sol lessivage
Mediterranean soils	Mediterraner Boden	Sol-mediterraneen
Mollisols		
Mor	Auflagehumus	Humus
Mull	Mull	Humus
Mull soils	Mull-Boden	Sol-a-mull
Muck	Niedermoor	Tourbe
Nitosols	Nitosol	
Noncalcic brown soils		
Ochrepts		
Orthents		
Orthids		
Orthods		
Orthox		
Oxisols		
Parachernozems	Smonitza	Chernozem sol d'alluvions
Paramosols	Paramosol	Sol-chatain
Pararendzina	Borowina	Rendzine sol d'alluvions

AMERICAN	GERMAN	FRENCH
Parasierozems	Parasierozem	Sierozem
Peat	Torf	Tourbe
Pelosols	Pelosol	Sol-hydromorphe
Pergelisols	Pergelisol	Sol-gele
Plaggenesch	Plaggenesch	Sol-riche-en-humus action-homme
Plaggepts		
Planosols	Planosol	Couche-rouge
Podzols	Podsol	Podzol
Podzoluvisols	Podzoluvisol	Sol lessivage
Poorly developed soils	Unreife-boden	Sol-peu-developpe
Prairie soils	Brunizem	Brunizem
Protopedons	Protopedon	Sol-brut or sol-d'alluvions
Psamments		
Pseudogleys	Pseudogley	Gley
Rambla	Rambla	Sol-brut or sol-d'alluvions
Ranker	Ranker	Ranker
Red desert soils	Roter Wuestenboden	Sol-subdesertique
Reddish brown soils	Roetlich-brauner Halbwueste Boden	Sol-subdesertique
Reddish-brown lateritic soils		
Reddish chestnut soils	Roetlich-Kastanienfarbiger Boden	Sol-chatain
Reddish prairie soils	Roetlicher Prairie Boden	Brunizem
Red-yellow podzolic soils	Gelbig roterpodsoliger Boden	Sol-ferralitique
Reg	Reg	Sol-de-desert
Regosols	Regosol	Sol-peu-evolue
Regur	Regur	Vertisol
Rendolls	.	
Rendzinas	Rendzina	Rendzine
Rhegosols	Rhegosol	Sol-peu-developpe
Rigosols	Rigosol	Sol-brut
Rutmark	Raamark	Sol-brut zone-froide
Saprists		
Sapropels	Sapropel	Sierozem
Sierozems	Sierozem	Sierozem
Solonchak soils	Solonchak	Sol-halomorphe
Solonetz soils	Solonetz	Sol-halomorphe
Soloth soils	Solod	Sol-halomorphe
Spodosols		
Stagnogleys	Stagnogley	Gley
Subarctic brown forest soils		
Subboreal desert soils		
Subboreal humid soils		
Subboreal steppe soils		
Subtropical desert soils		

AMERICAN	GERMAN	FRENCH
Subtropical dry soils		
Subtropical humid soils		
Syrogleys	Syrogley	Gley
Syrozems	Syrozem	Sol-brut zone-temperee
Takyr	Takyr	Sol-de-desert
Terrae calcis	Terrae calcis	Couche-rouge
Terra rossa	Terra rossa	Couche-rouge
Terra fusca	Terra fusca	Couche-rouge
Tir	Tir	Vertisols
Torrox		
Torrerts		
Tropepts		
Tropical desert soils		
Tropical dry soils		
Tropical humid soils		
Tundra soils	Tundra-Boden	Sol-de-toundra
Udalfs		
Uderts		
Udolls		
Udults		
Ultisols		
Umbrepts		
Ustalfs		
Ustserts		
Ustolls		
Ustsults		
Vega	Vega	Sol-brun or sol d'alluvions
Vertisols		
Wet meadow soils	Wiesenboden	Sol-hydromorphe
Xeralfs		
Xerests		
Xerolls		
Xerosols	Xerosol	Sierozem
Xerults		
Yeltozems	Yeltozem	Sol-mediterraneen
Yermosols	Ermosol	Sol-de-desert
Zonal soils	Zonaler Boden	Sol

N – Sediments

B.I.G. SEARCHING
Look under the appropriate second-level terms under the
heading sediments.

Other related headings include: sedimentary rocks,
sedimentary structures, sedimentation, soils, weathering,
and the individual commodities such as kaolin deposits
and clays.

---

sediments
  clastic sediments
    alluvium
    boulder clay
    boulders
    clay
    colluvium
    eluvium
    erratics
    flint clay
    gravel
    kaolin
    loess
    mud
    ooze
    pebbles
    proluvium
    residual clays
    residuum
    sand
    silt
    till
  carbonate sediments
  marine sediments
  organic residues

O – Geographic terms

INDEXING

Sets which have geographic terms on level 1 are called area sets. The terms in List O which are not followed by (3) are the only geographic terms permitted on level 1 in area sets. The possible level-2 and level-3 terms in area sets are given in List B.

Level 2 is never a geographic term.

While level-3 terms are usually topics, for area sets having either <u>areal geology</u> or <u>volcanology</u> on level 2, geographic terms may be used on level 3. The terms in List O marked with a (3) are valid only on level 3.

In certain sets other than area sets, geographic terms may be used on level 3. For example,

hydrology
  surveys
    name of river or
    lake or area
	paleogeography
	  age
	  area

ground water
  surveys
    name of aquifer
    or area
	clay mineralogy
	  areal studies
	  area

soils
  surveys
  area
	geodesy
	  surveys
	  area

In the above six examples, the geographic terms on level 3 must be valid on level 1 except where names of rivers, lakes, or aquifers are used.

B.I.G. SEARCHING

All terms on List O, except those designated (3), are headings in the Subject Index of the <u>Bibliography and Index of Geology</u>. It is our practice to make an area set (see List B, Area Sets) for each paper cited in GeoRef, excepting of course the relatively few geology papers which do not relate to geographic locations.

Note that the most specific term from List O which is applicable for a paper is used as the heading. Consequently a search for a broad region should include all terms in List O not designated (3), on that region. In each issue of the Bibliography, 'see also' references from broad terms to narrower terms used in that issue are included in the Subject Index, such as the following from the May 1981 issue:

Africa *see also* Algeria; Benin; Botswana; Cape Verde Islands; Djibouti; Egypt; Ethiopia;... Zambia

In the List which follows, the geographic terms are used in index sets on levels one and three, except those followed by (3) which are only permitted on level three.

global (3)
  Western World (3)
  - - - - - - - - -
  Eastern Hemisphere
  Northern Hemisphere
  Southern Hemisphere
  Western Hemisphere
  - - - - - - - - -
Africa
  Sahara
  East African Lakes (3)
    (Lake Albert, Lake Baringo,
    Lake Edward, Lake Kariba, Lake
    Kivu, Lake Magadi, Lake Malawi,
    Lake Natron, Lake Tanganyika,
    Lake Turkana, Lake Victoria) (3)
  - - - - - - - - - - - - -
  Central Africa (3)
    Angola
    Burundi
    Central African Republic
    Congo
    Equatorial Guinea
    Gabon
    Rwanda
    Zaire
      Bandundu (3)
      Equatorial Zaire (3)
      Kasai (3)
      Kivu (3)
      Lower Zaire (3)
      Shaba (3)
      Upper Zaire (3)
  East Africa (3)
    Djibouti
    Ethiopia
    Kenya
    Malawi
    Mozambique
    Uganda
    Somali Republic
    Sudan
    Tanzania
    Zambia
  North Africa (3)
    Algeria
      Ahaggar (3)
      Algiers region (3)
      Constantine region (3)
      Oasis region (3)
      Oran region (3)
      Saoura region (3)
    Egypt
      Eastern Desert (3)
      Middle Nile Valley (3)
      Nile Delta (3)
      Sinai (3)
      Western Desert (3)

Africa-Continued
  North Africa-Continued
    Libya
    Morocco
      Anti-Atlas (3)
      Ceuta (3)
      Melilla (3)
      Moroccan Atlas Mountains (3)
      Rif (3)
      Western Sahara
    Tunisia
  Southern Africa (3)
    Botswana
    Lesotho
    Namibia
    South Africa
      Cape Province (3)
      Natal (3)
      Orange Free State (3)
      Transvaal (3)
    Swaziland
    Zimbabwe
  West Africa (3)
    Benin
    Cameroon
    Chad
    Gambia
    Ghana
    Guinea
    Guinea-Bissau
    Ivory Coast
    Liberia
    Mali
    Mauritania
    Niger
    Nigeria
      East-Central Nigeria (3)
      Northern Nigeria (3)
      Southeastern Nigeria (3)
      West-Central Nigeria (3)
      Western Nigeria (3)
    Senegal
    Sierra Leone
    Togo
    Upper Volta

America (3)
  Caribbean region
  Latin America (3)
  - - - - - - - - - -
  Central America
    Belize
    Costa Rica
    El Salvador
    Guatemala
    Honduras
    Nicaragua
    Panama

America-Continued
  North America
    Appalachians
    Atlantic Coastal Plain
    Basin and Range Province
    Canadian Shield
    Colorado Plateau
    Columbia Plateau
    Great Basin
    Great Lakes
    Great Lakes region
    Great Plains
    Gulf Coastal Plain
    Mississippi Valley
    Rocky Mountains
    Western Interior
    - - - - - - - - - -
    Canada
      Eastern Canada (3)
        Maritime Provinces
          New Brunswick
          Nova Scotia
          Prince Edward Island
        Labrador
        Newfoundland
        Ontario
        Quebec
      Northwest Territories
        District of Franklin (3)
          Arctic Archipelago (3)
        District of Keewatin (3)
        District of Mackenzie (3)
      Western Canada (3)
        Alberta
        British Columbia
        Manitoba
        Saskatchewan
        Yukon Territory
    Mexico
    Saint Pierre and Miquelon
    United States
      Eastern U.S.
        New England
          Connecticut
          Maine
          Massachusetts
          New Hampshire
          Rhode Island
          Vermont
        Delaware
        District of Columbia
        Florida
        Georgia
        Maryland
        New Jersey
        New York
        North Carolina
        Pennsylvania

America-Continued
  North America-Continued
    United States-Continued
      Eastern U.S.-Continued
        South Carolina
        Virginia
        West Virginia
      Midwest
        Illinois
        Indiana
        Iowa
        Kansas
        Michigan
        Minnesota
        Missouri
        Nebraska
        North Dakota
        Ohio
        South Dakota
        Wisconsin
      Southern U.S.
        Alabama
        Arkansas
        Kentucky
        Louisiana
        Mississippi
        Tennessee
      Southwestern U.S.
        Arizona
        New Mexico
        Oklahoma
        Texas
      Western U.S.
        Pacific Coast
          California
          Oregon
          Washington
        Alaska
        Colorado
        Hawaii
        Idaho
        Montana
        Nevada
        Utah
        Wyoming

South America
  Andes
  - - - - - - - - - - - -Con - - - - - -
  Argentina
    Argentine Andes (3)
    Pampas (3)
    Patagonia (3)
  Bolivia
  Brazil
    Acre (3)
    Alagoas (3)
    Amapa (3)

America - Continued
  South America - Continued
    Brazil - Continued
      Amazonas (3)
      Bahia (3)
      Ceara (3)
      Espirito Santo (3)
      Goias (3)
      Maranhao (3)
      Mato Grosso (3)
      Minas Gerais (3)
      Para (3)
      Paraiba (3)
      Parana (3)
      Pernambuco (3)
      Piaui (3)
      Rio de Janeiro (3)
      Rio Grande do Norte (3)
      Rio Grande do Sul (3)
      Rondonia (3)
      Roraima (3)
      Santa Catarina (3)
      Sao Paulo (3)
      Sergipe (3)
    Chile
    Colombia
    Ecuador
    Falkland Islands
    Guyana
    French Guiana
    Paraguay
    Peru
    Surinam
    Uruguay
    Venezuela
      Apure Guarico Plain (3)
      Maracaibo Falcon Plain (3)
      Venezuelan Andes (3)
      Venezuelan Islands (3)
      Eastern Venezuela (3)
      Southeastern Venezuela (3)
  West Indies
    Bahamas
    Cayman Islands
    Greater Antilles
      Cuba
      Dominican Republic
      Haiti
      Jamaica
      Puerto Rico
    Lesser Antilles
      Barbados
      Dominica (3)
      Leeward Islands (3)
        Guadeloupe
        Virgin Islands (3)
          British Virgin Islands (3)
          U. S. Virgin Islands (3)

America - Continued
  West Indies - Continued
    Lesser Antilles - Continued
      Netherlands Antilles
      Trinidad and Tobago
      Windward Islands (3)
        Grenada (3)
        Martinique (3)
Asia
  Himalayas
  - - - - - - - - - - - -
  Arabian Peninsula
    Bahrain
    Kuwait
    Oman
    Qatar
    Saudi Arabia
    Southern Yemen
    United Arab Emirates
    Yemen
  Far East
    Borneo (3)
    Indochina
    - - - - - - - - - - - -
    Brunei (3)
    Burma
    Cambodia
    China
    Hong Kong
    Indonesia
      Celebes (3)
      Irian Jaya (3)
      Java (3)
      Kalimantan (3)
      Lesser Sunda Islands (3)
      Moluccas (3)
      Sumatra (3)
    Japan
      Hokkaido (3)
      Honshu (3)
      Kyushu (3)
      Ryukyu Islands (3)
      Shikoku (3)
    Korea
      North Korea (3)
      South Korea (3)
    Laos
    Malaysia
      Sabah (3)
      Sarawak (3)
    Mongolia
    Philippine Islands
    Singapore
    Taiwan
    Thailand
    Vietnam
  Indian Peninsula (3)
    Afghanistan

Asia-Continued
  Indian Peninsula-Continued
    Bangladesh
    Bhutan
    India
      Andhra Pradesh (3)
      Bengal Islands (3)
      Bihar (3)
      Gujarat (3)
      Haryana (3)
      Himachal Pradesh (3)
      Jammu and Kashmir (3)
      Karnataka (3)
      Kerala (3)
      Laccadive Islands (3)
      Madhya Pradesh (3)
      Maharashtra (3)
      Northeastern India (3)
      Orissa (3)
      Punjab State (3)
      Rajasthan (3)
      Tamil Nadu (3)
      Uttar Pradesh (3)
      West Bengal (3)
    Nepal
    Pakistan
    Sri Lanka
  Middle East
    Cyprus
    Iran
    Iraq
    Israel
    Jordan
    Lebanon
    Syria
    Turkey
      Anatolia (3)
      Pontic Mountains (3)
      Sea of Marmara region (3)
      Taurus Mountains (3)
      Turkish Aegean region (3)
Atlantic Ocean Islands
  Azores
  Bermuda
  Canary Islands
  Cape Verde Islands
  Madeira
  Sao Tome (3)
  Saint Helena (3)
Atlantic region
Australasia
  Australia
    New South Wales
    Northern Territory
    Queensland
    South Australia
    Tasmania

Australasia-Continued
  Australia-Continued
    Victoria
    Western Australia
  New Zealand
    North Island (3)
    South Island (3)
  Papua New Guinea
  East Pacific Ocean Islands
    Easter Island (3)
    Galapagos Islands
Eurasia
  Baltic region
Europe
  Alps
    Eastern Alps (3)
    Western Alps (3)
  Balkan Peninsula
  Carpathians
  Dinaric Alps (3)
  Jura Mountains (3)
  Pyrenees
  - - - - - - - - -
  Central Europe (3)
    Austria
      Austrian Vienna Basin (3)
      Central Austrian Alps (3)
      North Austrian Alps (3)
      North Austrian Crystallines (3)
      North Austrian Molasse (3)
      South Austrian Alps (3)
      South Austrian Molasse (3)
      - - - - - - - - - - - - - -
      Burgenland (3)
      Carinthia (3)
      Lower Austria (3)
      Salzburg State (3)
      Styria (3)
      Tyrol (3)
      Upper Austria (3)
      Vorarlberg (3)
    Czechoslovakia
      Berounka System (3)
      Bohemia-Moravia System (3)
      Bohemian Basin (3)
      Czech Bohemian Forest (3)
      Czech Erzgebirge (3)
      Czech Sudeten Mountains (3)
      Slovakia (3)
        Slovakian Carpathians (3)
        Slovakian Pannonian Basin (3)
    Germany
      Alpenvorland (3)
      Bavarian Alps (3)
      Bavarian Massif (3)
      Harz Mountains (3)
      Hesse Basin (3)

Europe-Continued
  Central Europe-Continued
    Germany-Continued
      Northeastern German Plain (3)
      Northwestern German Plain (3)
      Northern German Hills (3)
      Rhenish Schiefergebirge (3)
      Rhine Westphalian Basin (3)
      Saar-Nahe Basin (3)
      Saxonian Massif (3)
      Southwestern German Hills (3)
      Southwestern German Massifs (3)
      Thuringian Hills (3)
      Thuringian Massif (3)
      Upper Rhine Valley (3)
      - - - - - - - - - - - -
      East Germany
        Cottbus Bezirk (3)
        Dresden Bezirk (3)
        Erfurt Bezirk (3)
        Frankfurt Bezirk (3)
        Gera Bezirk (3)
        Halle Bezirk (3)
        Karl-Marx-Stadt Bezirk (3)
        Leipzig Bezirk (3)
        Magdeburg Bezirk (3)
        Neubrandenburg Bezirk (3)
        Potsdam Bezirk (3)
        Rostock Bezirk (3)
        Schwerin Bezirk (3)
        Suhl Bezirk (3)
      West Germany
        Baden-Wurttemberg (3)
        Bavaria (3)
        Hesse (3)
        Lower Saxony (3)
        North Rhine-Westphalia (3)
        Rhineland-Palatinate (3)
        Saarland (3)
        Schleswig-Holstein (3)
    Hungary
      Alfold (3)
      Budapest region (3)
      Central Transdanubia (3)
      Northeastern Hungarian Hills (3)
      Northwestern Transdanubia (3)
      Southeastern Transdanubia (3)
      Southern Transdanubia (3)
    Liechtenstein (3)
    Poland
      Northeastern Polish Plain (3)
      Northwestern Polish Plain (3)
      Polish Carpathians (3)
      Polish Sudeten Mountains (3)
      Southeastern Polish Hills (3)
      Swiety Krzyz Mountains (3)

Europe-Continued
  Central Europe-Continued
    Switzerland
      Central Swiss Alps (3)
      Eastern Swiss Alps (3)
      Northern Swiss Alps (3)
      Southern Swiss Alps (3)
      Swiss Jura Mountains (3)
      Swiss Rhine Graben (3)
      Swiss Molasse Basin (3)
  Southern Europe (3)
    Albania
    Bulgaria
      Balkan Mountains (3)
      Bulgarian Dobruja (3)
      Bulgarian Rhodope Mountains (3)
      Central Bulgaria (3)
      North Bulgarian Hills (3)
      West Bulgarian Hills (3)
    Greece
      Crete (3)
      Epirus (3)
      Euboea (3)
      Greek Aegean Islands (3)
      Greek Ionian Islands (3)
      Greek Macedonia (3)
      Greek Thrace (3)
      Peloponnesus (3)
      Sterea Ellas (3)
      Thessaly (3)
    Iberian Peninsula (3)
      Portugal
      Spain
        Asturian Arc (3)
        Betic Zone (3)
        Cantabrian region (3)
        Catalonian Coastal Ranges (3)
        Duero Basin (3)
        Ebro Basin (3)
        Galaico Massif (3)
        Guadalquivir Basin (3)
        Hercinico Centro (3)
        Hercinico Sur (3)
        Iberica (3)
        Spanish Pyrenees (3)
        Tagus Basin (3)
        - - - - - - - - - - - -
        Andalusia (3)
          Almeria Province (3)
          Cadiz Province (3)
          Cordoba Province (3)
          Granada Province (3)
          Huelva Province (3)
          Jaen Province (3)
          Malaga Province (3)
          Seville Province (3)

Europe-Continued
  Western Europe-Continued
    France-Continued
      French Pyrenees (3)
      Paris Basin (3)
      Saone-Rhone Basin (3)
      Vosges Mountains (3)
      - - - - - - - - - - - - - -
      Ain (3)
      Aisne (3)
      Allier (3)
      Alpes-de-Haute-Provence (3)
      Alpes-Maritimes (3)
      Ardeche (3)
      Ardennes Department (3)
      Ariege (3)
      Aube (3)
      Aude (3)
      Aveyron (3)
      Bas-Rhin (3)
      Belfort (3)
      Bouches-du-Rhone (3)
      Calvados (3)
      Cantal (3)
      Charente (3)
      Charente-Maritime (3)
      Cher (3)
      Correze (3)
      Corsica
      Cote-d'Or (3)
      Cotes-du-Nord (3)
      Creuse (3)
      Deux-Sevres (3)
      Dordogne (3)
      Doubs (3)
      Drome (3)
      Essonne (3)
      Eure (3)
      Eure-et-Loir (3)
      Finistere (3)
      Gard (3)
      Gers (3)
      Gironde (3)
      Haut-Rhin (3)
      Haute-Garonne (3)
      Haute-Loire (3)
      Hautes-Alpes (3)
      Haute-Marne (3)
      Haute-Saone (3)
      Haute-Savoie (3)
      Haute-Vienne (3)
      Hautes-Pyrenees (3)
      Hauts-de-Seine (3)
      Herault (3)
      Ille-et-Vilaine (3)

Europe-Continued
  Western Europe-Continued
    France-Continued
      Indre (3)
      Indre-et-Loire (3)
      Isere (3)
      Jura (3)
      Landes (3)
      Loir-et-Cher (3)
      Loire (3)
      Loire-Atlantique (3)
      Loiret (3)
      Lot (3)
      Lot-et-Garonne (3)
      Lozere (3)
      Maine-et-Loire (3)
      Manche (3)
      Marne (3)
      Mayenne (3)
      Meurthe-et-Moselle (3)
      Meuse (3)
      Morbihan (3)
      Moselle (3)
      Nievre (3)
      Nord Department (3)
      Oise (3)
      Orne (3)
      Paris (3)
      Pas-de-Calais (3)
      Puy-de-Dome (3)
      Pyrenees-Atlantiques (3)
      Pyrenees-Orientales (3)
      Rhone (3)
      Saone-et-Loire (3)
      Sarthe (3)
      Savoie (3)
      Seine-Maritime (3)
      Seine-et-Marne (3)
      Seine-Saint-Denis (3)
      Somme (3)
      Tarn (3)
      Tarn-et-Garonne (3)
      Val-de-Marne (3)
      Val-d'Oise (3)
      Var (3)
      Vaucluse (3)
      Vendee (3)
      Vienne (3)
      Vosges (3)
      Yonne (3)
      Yvelines (3)
    Iceland
    Ireland
    Luxembourg
    Netherlands

Europe-Continued
  Western Europe-Continued
    Scandinavia
     Denmark
      Faeroe Islands
      Finland
      Norway
       Northern Norway (3)
       Southern Norway (3)
      Sweden
       Gotland (3)
    United Kingdom
     English Channel Islands
     Great Britain
      England
       London Basin (3)
       Pennines (3)
       South-West England (3)
       Welsh Borderland (3)
      Scotland
       Grampian Highlands (3)
       Hebrides (3)
       Midland Valley (3)
       Northern Highlands (3)
       Orkney Islands (3)
       Shetland Islands
      Wales
      Isle of Man (3)
     Northern Ireland
  Indian Ocean Islands
   Comoro Islands
   Malagasy Republic
   Maldive Islands
   Mauritius
   Reunion
   Seychelles
  Malay Archipelago
   New Guinea
  Mediterranean region
  Oceania
   Melanesia
    Fiji
    New Caledonia
    New Hebrides
    Solomon Islands
   Micronesia
    Caroline Islands (3)
    Gilbert Islands (3)
    Mariana Islands
    Marshall Islands
    Nauru Island (3)
   Polynesia
    Cook Islands (3)
    French Polynesia (3)
     Marquesas Islands (3)

Oceania-Continued
  Polynesia-Continued
   French Polynesia-Continued
    Society Islands
    Tahiti
    Tuamotu Islands (3)
   Samoa
   Tokelau (3)
   Tonga
   Tuvalu (3)
Pacific region
Polar regions (3)
 Antarctica
  Kerguelen Islands (3)
  Scotia Sea Islands (3)
 Arctic region
  Greenland
  Jan Mayen
  Spitsbergen
Red Sea region
USSR
 Russian Republic (3)
 Siberia (3)
 Soviet Arctic (3)
 Ukraine (3)
- - - - - - - - - - - -
Caucasus
 Greater Caucasus (3)
 Lesser Caucasus (3)
 Northern Caucasus (3)
Central Asia (3)
 Caspian Depression (3)
 Central Kazakhstan (3)
 Eastern Kazakhstan (3)
 Fergana Basin (3)
 Karakum (3)
 Kopet-Dag Range (3)
 Kyzylkum (3)
 Pamirs (3)
 Tien Shan (3)
 Turgay Basin (3)
 Ustyurt (3)
Northeastern USSR (3)
 Amur region (3)
 Chukchi Peninsula (3)
 Dzhugdzhur region (3)
 Kolyma Uplift (3)
 New Siberian Islands (3)
 Sikhote-Alin Range (3)
 Stanovoy Range (3)
 Verkhoyansk region (3)
Russian Platform (3)
 Baltic Plain (3)
  Estonia (3)
  Latvia (3)

USSR-Continued
  Russian Platform-Continued
    Baltic Plain-Continued
      Lithuania (3)
      Byelorussia (3)
      Crimea (3)
      Dnieper-Donets Basin (3)
      Franz Joseph Land (3)
      Moscow-Pechora Syneclise (3)
      Soviet Carpathians (3)
      Soviet Fennoscandia (3)
      Timan Ridge (3)
      Transcarpathia (3)
      Ukrainian Shield (3)
      Ukrainian Syneclise (3)
      Voronezh-Volga Anteclise (3)
    Siberian fold belt (3)
      Baikal region (3)
      Eastern Sayan (3)
      Soviet Altai Mountains (3)
      Tuva (3)
      Western Sayan (3)
      Western Transbaikalia (3)
    Siberian Platform (3)
      Anabar Shield (3)
      Angara-Lena Basin (3)
      Aldan Shield (3)
      North Siberian Plain (3)
      Tunguska (3)
      Vilyuy Syneclise (3)
      Yakutia region (3)
      Yenisei Ridge (3)
    Soviet Pacific region (3)
      Kamchatka Peninsula (3)
      Koryak Range (3)
      Kuril Islands (3)
      Sakhalin (3)
    Urals (3)
      Central Urals (3)
      Mugodzhar Hills (3)
      Northern Urals (3)
      Novaya Zemlya (3)
      Polar Urals (3)
      Southern Urals (3)
    West Siberia (3)
      Kuznetsk Alatau (3)
      Minusinsk Basin (3)
      Severnaya Zemlya (3)
      Siberian Lowland (3)
      Taymyr (3)
world ocean (3)
  Antarctic Ocean
    Bellingshausen Sea (3)
    Mid-Antarctic Ridge (3)
    Ross Sea (3)

world ocean-continued
  Antarctic Ocean-Continued
    Scotia Sea (3)
    Weddell Sea (3)
  Arctic Ocean
    Baffin Bay (3)
    Barents Sea (3)
    Beaufort Sea (3)
    Chukchi Sea (3)
    East Siberian Sea (3)
    Greenland Sea (3)
    Kara Sea (3)
    Laptev Sea (3)
    Mid-Arctic Ocean Ridge (3)
    Norwegian Sea (3)
    White Sea (3)
  Atlantic Ocean
    North Atlantic (3)
      Azores-Gibraltar Ridge (3)
      Cape Verde Atlantic (3)
      European Atlantic (3)
        Baltic Sea
          Gulf of Bothnia (3)
          Gulf of Finland (3)
          Gulf of Riga (3)
        English Channel
        Bay of Biscay (3)
        Irish Sea
        North Sea
      Gulf of Guinea (3)
      Guyanese Atlantic (3)
      North American Atlantic (3)
        Caribbean Sea
        Gulf of Mexico
        Gulf of Saint Lawrence (3)
        Hudson Bay (3)
        Labrador Sea (3)
      North Atlantic Ridge (3)
    South Atlantic (3)
      South American Atlantic (3)
      South Atlantic Ridge (3)
      Southeast Atlantic (3)
  Caspian Sea
  Indian Ocean
    Andaman Sea (3)
    Arabian Sea
      Gulf of Aden
      Gulf of Oman (3)
      Persian Gulf
    Bay of Bengal (3)
    Carlsberg Ridge (3)
    Central Indian Ridge (3)
    Mozambique Channel (3)
    Ninetyeast Ridge (3)
    Red Sea

world ocean-continued
  Indian Ocean-Continued
    Red Sea-Continued
      Gulf of Aqaba (3)
      Gulf of Suez (3)
      Southeast Indian Ridge (3)
      Southwest Indian Ridge (3)
    Mediterranean Sea
      East Mediterranean (3)
        Adriatic Sea
        Aegean Sea
        Black Sea
        Ionian Sea
      West Mediterranean (3)
        Alboran Sea (3)
        Gulf of Lion (3)
        Tyrrhenian Sea
  Pacific Ocean
    East Pacific Rise (3)
    North American Pacific (3)
      Gulf of Alaska (3)
      Gulf of California
    South American Pacific (3)
      Carnegie Ridge (3)
      Chile Ridge (3)
      Cocos Ridge (3)
      Nazca Ridge (3)
    South Polynesian Pacific (3)
    West Pacific (3)
      Bering Sea
      Coral Sea
      East China Sea
      Indonesian Seas (3)(Celebes Sea, Java Sea (3))
      Japan Sea
      North Australian Seas (3)(Arafura Sea (3),Timor Sea (3))
      Okhotsk Sea
      Philippine Sea
      South China Sea
        Gulf of Siam (3)
        Gulf of Tonkin (3)
      Tasman Sea
      Yellow Sea
    Pacific-Antarctic Ridge (3)

INDEX